Global Events:
TIPPING POINTS
Media and Communication

Dr. Philip Gordon, PhD

February, 2013
Paris, France
Blue Matrix Publications

Global Events: TIPPING POINTS

Media and Communication

Copyright © 2013, Dr. Philip Gordon, PhD

All Rights Reserved

USA:

2995 Woodside Rd. Suite 400

Woodside, CA 94062

France:

2 rue Despaty

Voisines, FR 89260

ISBN-13:978-1481261869

ISBN-10:148126186X

Typography: David Frith

www.MDG-Consultants.com

Printed in the United States of America

Dedication

To Jaike and Matthew

C urrent history offers a number of interesting case studies of *global events* and their relationships with *tipping points*. Each involves significant technological developments related to *media* communication and the impact on our world economically, culturally and politically.

The three case studies examined herein were selected on the basis of two main requirements—

Each is characterized by a defined *"moment in time,"* where circumstances and *events* change as a result of the *media* influences or *"phenomenon"* that could only be characterized as a tipping point, and

Each opened doors to new perspectives relative to *global events* and *media* communication and the creation of our new world reality.

Acknowledgements

My sincere gratitude and appreciation to the personnel at some of the libraries and archives where I consulted texts and documents in United States, France and Great Britain, who were particularly helpful in facilitating my access to relevant information. In this regard, I would like to single out the libraries at the *Centre d'etudes Diplomatiques et Strategiques*, Paris, France (PhD – 2012), *Johns Hopkins University* in Baltimore, Maryland (MS-1979) and *Sciences Po, L'Institut d'études Politiques (IEP) de Paris*.

This acknowledges that a book on this topic would have been impractical for a sole person to research efficiently, but with the emergence of the Internet and the proliferation of material available through it, has allowed me to obtain texts and documents from governments, academic institutions, publications and libraries. Occasionally, search results yielded information that proved relevant and valuable from sources that one would not normally associate with the specific topics or locations being examined; this only broadened my perceptions of the issues at hand, and I extend my gratitude to *people* I do not know but who brought the technology to market where it could be utilized for this project.

Finally, I would like to thank my family and friends for the encouragement they lent in connection with my preparation of this book. Especially my wife, Anne-Lyse and my sons Jaike and Matthew, who were patient with me as it took time away from them and my business to do this research. I am very grateful for their presence and understanding.

TABLE OF CONTENTS

Part One: Tipping Point Theory and Global Media

TABLE OF CONTENTS [*CONTINUED*]

TABLE OF CONTENTS [*CONTINUED*]

Bibliography

List of Charts, Tables and Figures

TABLE OF CONTENTS [*CONTINUED*]

TABLE OF CONTENTS [*CONTINUED*]

Foreword

FOREWORD

This book addresses three comparative case studies in which the concept of the *Global Events* and *Tipping Points* are evidenced and in many ways influenced by the *media*. The case studies were selected on the basis of three (3) common key *attributes*: *contagiousness, stickiness* (during their development), and *one dramatic moment* in time that could be defined as *Tipping Points*.

These case studies were also chosen largely because their *attributes* included the four (4) *characteristics* of: *people, organizations, media* and *events* that could be categorically analyzed and compared within a conceptual framework, including recent *events* that were impacted by *global media*.

The case studies are about the literature, trends, and information relative to *media impact* on *global events* in the context of the *Tipping Point Theory phenomenon*, which are interpreted in the context of much broader event themes. These case studies represent only a few of the *global events* that meet the criteria for the *Tipping Point Theory* analysis.

The first case study, the *Presidential Campaign of Barack Obama*, was chosen in order to examine a narrower scope and timeframe for the analysis with a finite endpoint, and also because geographically, politically, economically, and culturally, only one nation was involved. In contrast, the second case study, the *International Financial Crisis of 2007 through 2010*, involves more complex issues that impact other nation states and global economies. It requires an analysis on a much broader scale in time, (2007 through 2010), with its origins centered primarily in one country, yet, in contrast to the first study, represented a larger field of data and information to analyze.

The last case study, *Climate Change*, is included because, as the two first case studies provide a basis for understanding critical *tipping point attributes* and their *characteristics*, with its ongoing nature, the application to the

Climate Change presents probable solutions and parameters for analyzing how *tipping points* actually evolve and what their subsequent impact on a global scale could be.

The *Climate Change* case study involves many nation states and many complex contemporary *media* development issues and technology changes integral to the research that is continuing to evolve. The correlations between these *global events* and *media impact* exhibits very strong "*cause and effect*" relationships between the *tipping point attributes* and the case study sequential evolutions.

The case studies were also selected to be examined in this book on the basis of quantity and accessibility of information about the *attributes* and *characteristics*, their history in the context of *media* and impact, and emergence of a model theory framework to apply to future *global events*.

These three case studies encompass relevant examples with regard to *media impact* and *global events*. They offer considerable variety in terms of their geographic locations as well as their scale, historical, social and political context, and consequently, they lend themselves to comparisons that exhibit these similarities and differences. In the course of analyzing them, consideration is given to the subdominant questions of:

- *Why media has such an impact on global events?*
- *And why we have not already embraced this conceptual framework model for the purposes of addressing Climate Change?*

Throughout this book, many of the interpretations of the material are my own unless otherwise specified. References in the case studies are identified in alphabetical order so as to present a neutral and consistent way of referring to the sources in this context and therefore not implying any political or personal biases based on the order in which the references are named. Similarly, the text and information reproduced in the annexes are done alphabetically first, as they do not necessarily illustrate the timeline continuity of the *events* as they unfold during the case studies.

Other *global events* in retrospect might have been chosen, such as the recent *events* involving Syria, Tunisia, Egypt, Libya, Bahrain, and other countries in the Arab Spring regions. But due to time constraints and limited knowledge of the culture and language, these, and other more recent *events* were dismissed for future versions.

Part I

Tipping Point Theory and Global Media

INTRODUCTION

The *Tipping Point Theory* is a conceptual model framework application for analyzing *phenomena*, particularly in the context of *media* and *global events*, which helps to provide a strategy for international relations by collectively playing an important role in the interactions internally and between nation states.

The *Tipping Point Theory* model framework discussed in this book provides a means for analyzing and comparing *global events* that are impacted by various *media* types that arise and cross borders of nations and states that coexist and are codependent within the international system. The interpretation and adoption of the *Tipping Point Theory* provides a potential opportunity to reduce the conflict among the world community in the future, since demand for resources that are only now becoming apparent, does not match future political, social and/or economic needs.

Part one of this version, first reviews the concept of the *Tipping Point Theory* that is fundamental to our analysis, defining and characterizing the *attributes* and *characteristics*. It then examines, presents and analyzes the academic literature on the subject of *media* and its global impact as relevant from information sources that are directly involved with their creation and delivery. It attempts to explain the *media impact* from the perspective of its development and evolution in an international context and then discusses their relationships and interactions that exist within the medium and the world today and with *events* that happened in the past and future.

While there are many published works about the broader concepts of *media impact* and *global events*, by the time one focuses on the narrower *attributes* and *characteristics* and concepts of our *Tipping Point Theory*, the literature is fairly limited and almost none of it is relevant or recent.

Additional research has recently been spawned by contemporary *events* (for example, the diplomatic cable discoveries from *WikiLeaks* and deployment via the Internet and their effect on nation state diplomacy); however, when the analysts discuss their impacts, they tend to summarize by providing a rather limited viewpoint based on previous published approaches and in doing so, they confirm how little has gone forth towards analyzing what impact the *media* really has on *global events*.

One development that emerges from this process is that nearly all of the underlying concepts that are critical to the analysis of *media impact* and *global events* have been subject to varying definitions, divergent theories and conflicting interpretations. It thus became essential to develop a workable definition of the *Tipping Point Theory* and provide a perspective for the *attributes* and *characteristics* by which one can synthesize the information prevalent in the literature, and thereby construct a solid framework from which to proceed with the subsequent examination of the *attributes* and *characteristics* in each of the case studies that form the core of this book.

Another highlight is that the *media impact* on *global events*, in terms of its analysis, when viewed in the context of contemporary *media* issues, appears to be more "*ad hoc*," without a systematic process for problem-solving, forecasting and guidance for international management. This illustrates that the worlds' nations gravitate towards a more pragmatic behavioral approach in their international event solutions, defying logic, and in doing so, often create additional problems, without focusing on the issues at hand.

The effort to understand *media impact* on *global events* presents challenges and highlights the very nature and essence of the *Tipping Point Theory* and the concepts discussed here, particularly, anticipating when a tipping point will occur means you have to develop an approach to systematically identify it first, and then perhaps, deploy a "*solutions methodology*" to predict (or avoid) it on a extremely large global scale (when it is in fact detrimental to mankind).

The analysis of the *Tipping Point Theory* with regard to *media* impacts on *global events* illustrates the huge gaps that exist in the application of this conceptual model. As such, while the application this theory and model - by my own admission, might be considered on the periphery of international relations, it does, in fact, find itself centered in the middle of the most, (if not all), of our most pressing global arguments.

For purposes of organization, *Chapter One* presents the definition of the *Tipping Point Theory* and the *attributes* and *characteristics*, while *Chapter Two* overviews Global *Media*, highlighting development and key relationships.

CHAPTER 1: THE TIPPING POINT THEORY

Section 1.1: Tipping Point Theory

The *Tipping Point Theory* is defined in this context as the "*biography of an idea,*" where the idea can be very simple. It is an approach to understand the emergence of a *trend*, or the ebb and flow of a trend.

For examples, the election of the first bi-racial US president, or the meltdown during the international financial crisis; or the emergence of global climate awareness. These *phenomenon* of word of mouth, or any number of the other mysterious changes that influence everyday life and their relative impacts, is to think of them in this context as *epidemics* - ideas, products, messages and behaviors as explored in this context *spread just like viruses* do.

Three *attributes* of The *Tipping Point Theory* are to be explored in this context of *media impact* and *global events*:

First, the relationship to *contagiousness*, or "*spread of linkages*" in the *characteristics*; two, *stickiness*, the fact that "*little causes can have big effects*"; and three, that change happens not gradually but at "*one dramatic moment.*" In fact this third trait—supporting the notion that *phenomena* of *epidemics* can rise and fall in *one dramatic moment*—is the most important of the three *attributes* because it is this one that makes sense of the first two and perhaps permits the greatest insights into why *global events* happen the way they do with *media* influence.

This definition given to that "*one dramatic moment*" when during the *epidemic* everything can change all at once, and is termed in this context as, the "*Tipping Point.*"[1] As *global events* evolve, the world is impacted in terms of the *media* delivery systems, population response and the way countries are governed. *Global events* provide a context for the case study analysis of the *media* in context of these three (3) tipping point *attributes*. These collectively bring a sense of relevancy if viewed from the dual perspective of international relations and the world's future.

Tipping Point Theory attributes are further analyzed with four (4) *characteristics*:

People, Organizations, Media and Events

Each of these *characteristics* presents dynamic relationships on how they relate to the sequential *attributes* and the others in this context. The *characteristics* were captured in the research aiming at those most salient to the discussion and key questions regarding the existent of the *phenomena* of a tipping point. These *characteristics* are initially defined as:

- *People*: key individuals central to the studies and *attribute* progression

- *Organizations*: pivotal in their roles and responsibilities throughout the studies

- *Media*: types, audience, technology, participation and influence

- *Events*: global milestones, outcomes and moments reinforcing the underlying theory premises

Moreover, each of these *characteristics* evolved during different "*windows*" or, duration of periods of time for each of the case studies that are cited in this book. Government, corporation, and private individual intervention with the *media* at any point in history are relevant in the discussion and become part of the argument as to how *global events* are in fact impacted and influenced. The *Tipping Point Theory* illustrates *a conceptual model framework* approach for analyzing *media impact and global events*.

What emerges from a review of the literature about this theory is that, while the concept is loosely defined, there exist significant variations on how it can be interpreted with regard to global event *phenomena*.[2]

---

1   Malcolm GLADWELL, *The Tipping Point: How Little Things can Make a Big Difference*, New York: Back Bay Books, 2002.
2   Ibid.

Several French tipping point, "*point de bascule*" theories have emerged over the years offering similar interpretations. One in particular, the *Catastrophe Theory*, which originated with the work of the French mathematician René Thom in the 1960s and then later expanded by an American theorist Tim Poston in 1990s.[3] Which proposed that small changes in certain parameters of a nonlinear system can cause "*equilibria*" to appear or disappear, or to change from attracting to repelling and vice versa, leading to large and sudden changes, or *degenerate points* of the behavior of the system, i.e. *Tipping Points*.

The French Catastrophe Theory further points out that these *degenerate points* are not merely accidental, but are *structurally stable,* until they degenerate, or are influenced enough so that they become in fact *tipping points*.

What follows is a more detailed definition of the *Tipping Point Theory*, its *attributes* and *characteristics* and their relationships that are evident in our case studies regarding *global events*.

Section 1.2: Tipping Point Attributes

Section 1.2.1: Contagiousness Attribute

We first explore the *contagiousness attribute* in context with the *Tipping Point Theory* applied as it is characterized by the *Broken Windows Theory*. *Broken Windows* was the brainchild of James Q. Wilson and George Kelling. Wilson and Kelling argued that in effect impacts and events are the inevitable result of *disorder*.[4]

A simple application of this *phenomenon* suggests that if a window is broken and left unrepaired, i.e. the international financial system, whereby if it appears that the system is unregulated, unmanaged, or that no one cares about it, and no one is in charge, soon more windows will be broken, and the condition will spread sending a signal that anything goes. In countries or regions illustrating this example, financial oversights and lack of monitoring regulation are all the equivalent of *broken windows* and promote invitations to more serious financial meltdowns as is evident in events in Ireland and Greece that affected the entire European Union.[5]

---

3   Poston, Tim and Stewart, Ian. Catastrophe: Theory and Its Applications. New York: Dover, 1998.
4   James WILSON and George KELLING, "Broken Windows. The Police and Neighbourhood Safety," *Atlantic Magazine*, 3, 1982.
5   Ibid.

This is the *epidemic of finance*. Its effects are *contagious* — just as a *trend* is *contagious*—in that it can start with just one *broken window*, or country bank failure, and *spread* instantaneously, particularly if the message is delivered via television or internet to an entire region's population in the world economies. The impetus to engage in a certain type of behavior is not coming from a particular kind of person or culture but from a feature of the environment, i.e. financial sector–banking. *Contagiousness* postulates that if a particular behavior in a community (or world) goes unaddressed, it signals that nobody cares about the community (or world) resulting in additional behavior of the same type.

The *contagiousness attribute* proposes that there are *people, organizations,* and *events* within the global *media* environments that are capable of initiating these *epidemics* by starting with small, seemingly inconsequential causes that ultimately have large (global) impacts.

Broken Windows Theory in this context, viewed from the Malcolm Gladwell interpretation states: "*The success of any kind of social epidemic is heavily dependent on the involvement of (three types of) people with a particular and rare set of social gifts.*" According to Gladwell, economists call this the 80/20 Principle, which is the idea that in any situation roughly 80 percent of the "*work*" will be done by 20 percent of the participants.[6] These *people* are described in the following ways:

People who "*link us up with the world...people with a special gift for bringing the world together.*" They are "*a handful of people with a truly extraordinary knack [...for] making friends and acquaintances.*" He characterizes these individuals as having comprised social networks of over one hundred people. To illustrate, Gladwell cites a number of examples: the midnight ride of Paul Revere; Milgram's experiments in the *small world* problem; the "*Six Degrees of Kevin Bacon*" trivia game; Dallas businessman Roger Horchow; and Chicagoan Lois Weisberg, a person who understands the concept of the weak tie.

Gladwell attributes the social success of connectors to "*their ability to span many different worlds [...as] a function of something intrinsic to their personality, some combination of curiosity, self-confidence, sociability, and energy.*"[7]

---

6   Malcolm GLADWELL, The Tipping Point, op. cit.
7   Ibid.

Information specialists, or *"people we rely upon to connect us with new information."* They accumulate knowledge, especially about the marketplace, and know how to share it with others.[8] Gladwell cites Mark Alpert as a prototypical maven who is *"almost pathologically helpful,"* further adding, *"He can't help himself."* In this vein, Alpert himself concedes, *"A maven is someone who wants to solve other people's problems, generally by solving his own."* According to Gladwell, mavens start *"word-of-mouth epidemics"* due to their knowledge, social skills, and ability to communicate. As Gladwell states, *"Mavens are really information brokers, sharing and trading what they know."*[9]

Finally, *persuaders,* charismatic people with powerful negotiation skills. They tend to have an indefinable trait that goes beyond what they say, which makes others want to agree with them. Gladwell's examples include California businessman Tom Gau and news anchor Peter Jennings, and he cites several studies about the persuasive implications of non-verbal cues, including a headphone nod study (Wells and Petty)[10] and William Condon's cultural micro rhythms study.[11]

Section 1.2.2: Stickiness Attribute

The second *attribute* to be addressed is termed *stickiness.* In *The Tipping Point* Gladwell identified traits that make ideas *sticky.* I take this *attribute* beyond the scope of Gladwell's book. Gladwell was interested in what makes social epidemics *"epidemic."* The focus here is how particular *phenomena* are constructed and what makes some *stick* and others disappear in the context of *media,* particularly the *global events* highlighted in the case studies.

What I am looking for here are the same *attributes,* the same *characteristics* that are reflected in a wide range of case study applications. In general there seem to be *six principles* at work in order to make *phenomenon* sticky and these are:

- *Simplicity,* the golden rule – it has to be simple and profound.

- *Unexpectedness,* it must generate interest and curiosity.

- *Concreteness,* the only way to reach everyone in the audience.

- *Credibility,* the credentials need to speak for themselves.

---

8 Ibid.
9 Ibid.
10 Gary WELLS and Richard PETTY, "The Effects of Head Movement on Persuasion: Compatibility and Incompatibility of Responses," *Basic and Applied Social Psychology*, 1, 1980, 219–230.
11 William CONDON, "Cultural Microrhythms," in *Interaction Rhythms*, 1974, New York, Human Sciences.

- *Emotions*, people must feel something.

- *Stories*, stimulating people to respond quickly.

I explore each these *principles* as further with the four (4) *Tipping Point Theory characteristics* of: *people, organizations, media and events.*

On the other hand, the *stickiness attribute* is rather straightforward; there is a simple way to package information such as the Internet that under the right circumstances can make it irresistible. And all you have to do is find it. The presidential campaigns of Barack Obama could be mostly characterized by this *attribute*, and the result correlated with available *media* technological tools he or his staff had at their disposal, and their methods and strategy for deployment.

Another example of "*little causes have big effects*" is where a specific content of a message renders its impact memorable: Popular children's television programs such as *Sesame Street* and *Blue's Clues,* pioneered the properties of *stickiness*, thus enhancing the effective retention of the educational content in parallel with its entertainment value.

Section 1.2.3: One Dramatic Moment Attribute

The third and last *attribute*, that change happens *not gradually* but at *one dramatic moment*, supports the notion that *epidemics* can be created and influenced to end in *one dramatic moment* – this is the most important of the three *attributes* because this one makes sense of the first two and perhaps permits the greatest insights into why modern *events* happen the way they do. The *attribute* given to that *one dramatic moment* in the *epidemic* when everything can change all at once is called the *tipping point* or is termed: "*Something Else.*"[12]

Based on these three *attributes* and the potential developments now exponentially gaining in world global *media* technology and also being readied for deployment in the not-too-distance future, one can correlate this theory with a series of world impacting *tipping points* happening very soon. "*Something else is when human behavior is sensitive to and strongly influenced by its environment,*" as Gladwell says, "*Epidemics are sensitive to the conditions and circumstances of the times and places in which they occur.*"

---

12   Malcolm GLADWELL, The Tipping Point, op. cit.

For example, "*zero tolerance*" efforts to combat minor crimes such as beating subway fares and vandalism on the New York subway that resulted in a decline in more violent crimes city-wide. Gladwell describes the "*bystander effect*," and explains how *Dunbar's number* plays into the tipping point, using Rebecca Wells' novel *Divine Secrets of the Ya-Ya Sisterhood*, evangelist John Wesley, and the high-tech firm *Gore Associates*.[13]

Tipping points are "*the levels at which the momentum for change becomes unstoppable.*" Gladwell defines a tipping point as, "*the moment of critical mass, the threshold, the boiling point.*" Gladwell attempts to explain and describe the "*mysterious sociological changes that mark everyday life.*" As Gladwell states, "*Ideas and products and messages and behaviors spread like viruses do.*" Other examples of such "*changes*" in his book include the rise in popularity and sales of *Hush Puppies* shoes in the mid-1990s and the precipitous drop in the New York City crime rate after 1990.[14]

Section 1.3: Tipping Points and Global Events

The tipping point can be seen in a variety of global developments, for example, among them is population change.

It is estimated that on August 26, 2011 the world's population reached 7 billion. And by August 10, 2045 the world's population is estimated to reach 9 billion (UN estimates). One could ask, in other words, *when is the world population going to reach a tipping point?* When will there be a scarcity of air, land, water, food, etc. in order to support the demands of societies and life-sustaining systems? [15]

The world's population, three centuries from now are projected to stabilize at 9 billion if fertility levels continue their decline, particularly in the developing world, but could also top more than 1.3 trillion if they remain unchanged from current rates according to statistics released by the United Nations. According to medium-level projections, women in every country will each have about two children in the decades to come, raising the world population from its current 6.4 billion to 9 billion in 2300 according to *UN*'s Population Division.

---

13 Ibid.
14 Ibid.
15 Pentti LINKOLA, (1992). "The Doctrine of Survival and Doctor Ethics." Paints the prospective options by the worlds' population when the tipping point is reached in her book. "What to do when a ship carrying a hundred passengers suddenly capsizes and there is only one lifeboat? When the lifeboat is full, those who hate life will try to load it with more people and sink the lot. Those who love and respect life will take the ship's axe and sever the extra hands that cling to the sides."

The following is a list of milestones regarding population change and a graphic depicting the projections in a conservative sense:

- 3 Billion: 30 January 1960
- 4 Billion: 8 September 1974
- 5 Billion: 31 March 1987
- 6 Billion: 20 January 1999
- 7 Billion: 26 August 2011
- 8 Billion: 27 April 2025
- 9 Billion: 10 August 2045

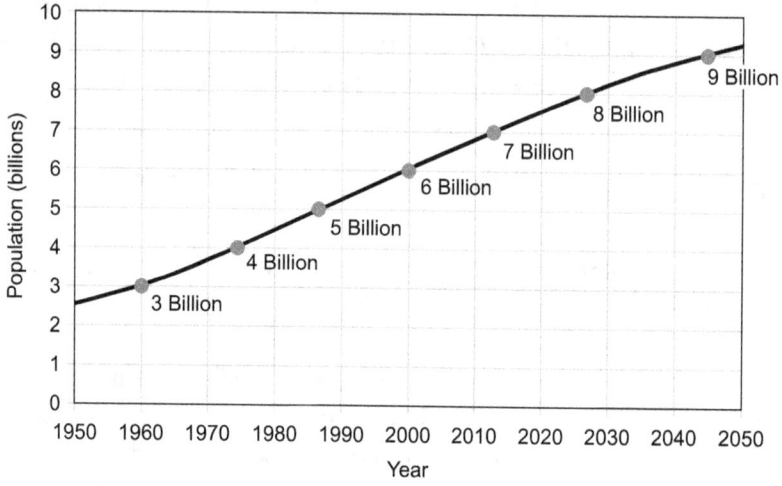

Source: U.S. Census Bureau, International Data base, December 2010 Update.

Figure 1.1: World Population 1950-2050[16]

Small variations in these forecasts will have enormous impacts in the long term, or a change such as one-quarter of a child under the two-child norm, or one-quarter of a child above the norm, would result in world populations ranging from 2.3 billion to 36.4 billion. If fertility levels remain unchanged at today's levels, however, world population would rise to 44 billion in 2100, 244 billion persons in 2150 and 1.34 trillion in 2300, according to the Division's new report, *World Population to 2300*. The UN said this clearly indicates that *"current high fertility levels cannot continue over the long term."*[17]

---

16   U.S. Census Bureau, International Database, June 2011 Update, World Population 1950–2050, 2010.
17   United Nations Department of Economic and Social Affairs/Population Division (UNDES), "World Population in 2300 to be around Nine Billion Persons," December 9, 2003.

Given progress in extending life expectancy, people could expect, on average, to live more than 95 years by 2300. Japan, which is the global leader in life expectancy today, is projected to have a life expectancy of more than 106 years by 2300.[18] In today's terms these figures alone present some startling statistics. Even at 7 billion, which is the population to be reached in August 2011, there is a growing awareness that the planet that we occupy is going through some significant changes that are either caused or aggravated by the existence and activities of mankind. With finite resources for land, water, food, and more importantly air, the additional influx and growth of population reaching anywhere between 9 billion to 44 billion and beyond presents some interesting analytical scenarios for a wide magnitude and range of potential *tipping points* that may occur as a result of the sheer need and scarcity of resources necessary for mankind for many of the basic requirements for life.

The example above is cited and described primarily to point out the magnitude and complexity of scaling and defining the parameters for any global *event* and analysis for *Tipping Point Theory* case studies. One would hope that with the above example we can apply our methodology (*conceptual model framework*) to identify the *attributes* of *contagiousness, stickiness,* and *one dramatic moment* that would predict that *tipping point* for the world relative to population. We would need to make assumptions about the *characteristics* for each, particularly the *people, organizations,* the *media* systems, and the sequential *events* that would lead up to the *tipping point.*

Our intent here is not to address population growth and to develop solutions for this challenge, albeit a worthy endeavor. Rather the intent is to point out the breadth and width and potential application of this theory.

Section 1.4: Tipping Points and Media Impacts

Media technologies have been integral to the representation of *global events.* In the current context, for example, the USA-led *Iraqi Freedom War* and the *War on Terror* were by most accounts of global international communication represented by news *media*, ignoring other genres, often neglected the crucial roles that audiences and newer technologies actually had within the "*real time*" global *media* culture.

The war in Iraq was one of the first major *media* events in which the technological infrastructure of the *Internet Age* facilitated the immediate and

18 Ibid.

global dissemination of; images; video; text and multimedia from journalists, civilians, and combatants via the Internet, Web, and mobile satellite networks that reached a worldwide audience, whereby both first and third world countries were exposed. (For example, representations of Saddam Hussein's execution could not be *'controlled'* by the Bush administration, as a video of his hanging was leaked and shown on television and over the Web).

It would be difficult to describe the degree to which the developments of the Internet has complicated the U.S. government's efforts to manage information surrounding war operations. The Bush administration no doubt tried to prevent all images of suffering and death occurring during the 2003 Iraq War (as was done during the 1991 Gulf War), with rare exceptions.

In response to a new global communications environment, the Pentagon has reportedly tried implementing electronic warfare (EW) strategies as outlined in the declassified *Information Operations Roadmap* commissioned by Donald Rumsfeld, past US Secretary of Defense in 2003. Approaching the Internet as an enemy weapon and deployment system, the *Roadmap's* objective is to "*[p]rovide a future EW capability sufficient to provide maximum control of the entire electromagnetic spectrum, denying, degrading, disrupting, or destroying the full spectrum of globally emerging communication systems, sensors, and weapons systems dependant on the electromagnetic spectrum.*"[19] A major focus was the ability to disable cell phones, which, was presented by the Pentagon and journalists as a sign of "*progress*" in Iraq. Particularly, because they are commonly used as remote detonators for roadside bombs, or IEDs, activated by Internet messaging applications. The *Roadmap* called for "*improvements in the capability ...to rapidly generate audience specific 'tipping points', commercial-quality products into denied areas*" and "*project ...electronic attacks into denied areas by means of stealthy platforms.*"[20] Concluding that the actual goal of the information war on terror is to remove or control all non-U.S. communication and news from designated regions, suggesting a reductive model of media imperialism as we examine this technology. The expansion of telecommunications and technology in the world has meant that, in effect, the world has gotten more deeply connected and become "*flat,*" as in Thomas Friedman's famous formulation.[21]

In today's world (2013) cheap phone calls and broadband have made it possible for people to communicate instantly around the globe. Historically,

---

19 National Security Agency (NSA), *Information Operations Roadmap* (National Security Archive Electronic Briefing Book No. 177), October 30, 2003.
20 Ibid.
21 Thomas FRIEDMAN, *The World is Flat: A Brief History of the Twenty-first Century*, New York, Farrar, Straus and Giroux, 2005.

with the arrival of the big ships in the 15ᵗʰ century, only goods could become mobile. With modern banking in the 17ᵗʰ century, capital then became mobile. Later in the 1900's labor became mobile with vast movements of populations searching for work opportunities. Similarly since the 1980's, three forces - politics, economics and technology - have pushed in the same direction to produce a more open, connected, exacting international *media* environment. These forces have given countries everywhere fresh opportunities to challenge positions on the world ladder, some for growth and prosperity goals, some not so admirable objectives. And now stretching into 2013, ideas are indeed more mobile with the technology.

Today people from all over the world have more access and are becoming more comfortable putting their own indigenous imprint on the world stage. "*Local*" and "*modern*" is now about how to communicate and co-exist side-by-side with global and Western. The same is true of foreign policies, world conflict, and *events*.

But there are some underlying realities. For example, basic issues of security influencing the global and immediate world neighborhood are critical components of any country's national security policy. And now with global technology reaching throughout the world so easily, no civilization can develop in a "*hermetic box*," which is apparent in the Arab region development of 2011–13. And importantly when it comes to religion and world views, countries are emerging from the backgrounds with rapid delivery and advanced security systems. These dynamics continue to present internal challenges overlaid with outside influences.

Section 1.5: Summary

As discussed, Gladwell describes the three rules of *epidemics* (*attributes*) in *tipping points* as:

- *Contagiousness*: the "*Law of the Few*"
- *Stickiness*: "*Little Causes have Big Effects*"
- *One dramatic moment*: "*Something Else*,"[22]

Whereas, these *attributes* have also been extended and applied in many fields, from the various sciences and economics to human ecology to epidemiology:

---

22 Malcolm GLADWELL, The Tipping Point, op. cit.

In *sociology*, a *tipping point* (or angle of repose) is the event of a previously rare phenomenon becoming rapidly and dramatically more common. The phrase was coined in its sociological use by Morton Grodzins by analogy with the fact in physics that *"adding a small amount of weight to a balanced object can cause it to suddenly and completely topple."* Grodzins studied integrating American neighborhoods in the early 1960s. He discovered that most of the white families remained in the neighborhood as long as the comparative number of black families remained very small. But at a certain point when *"one too many"* black families arrived, the remaining white families would move out en masse in a process known as white flight. He called that moment the *"tipping point."*[23] The idea was expanded and built upon by Nobel Prize-winner Thomas Schelling in 1972.[24] A similar idea underlies Mark Granovetter's threshold model of collective behavior.[25]

In *climate science* it has also describes the propagation and change of populations in an unbalanced ecosystem.

In *mathematics*, as the angle of repose is seen as an *inflection point*.

While in *control theory*, the concept of positive feedback describes the same *phenomenon*, with the problem of balancing an inverted pendulum being the classic embodiment.[26]

And finally, the tipping point in *physics* is the point at which an object is displaced from a state of stable equilibrium into a new equilibrium state qualitatively dissimilar from the first.

The discussion that follows, is premised on the development of a research and analysis methodology, utilizing, in fact, aspects of many of these interdisciplinary definitions, however, rooted predominantly of the Gladwell viewpoint and expanded here.

Before proceeding to the *attributes* and *characteristics* of each of the case studies, we move forward to *Chapter Two* with a review of *global media*.

---

23 Morton GRODZINS, *The Metropolitan Area as a Racial Problem*, Pittsburgh, University of Pittsburgh Press, 1958.
24 Thomas SCHELLING, "Dynamic Models of Segregation," *Journal of Mathematical Sociology*, 1, 1972, pp. 143–186.
25 Mark GRANOVETTER, "Threshold Model of Collective Behavior," *The American Journal of Sociology*, 83, 1978, pp. 1420–1443.
26 Malcolm GLADWELL, The Tipping Point, op. cit.

CHAPTER 2: GLOBAL MEDIA

Frrom radio, film, television, to the Internet and mobile satellite networks, *global media* communication technologies have been integral to the *Tipping Point phenomena* and representation of world *events*. In this chapter we explore the way the *media* has always been eager to improve communication delivery technologies and how countries have developed new communication technologies and *media* since at least the 19ᵗʰ century.

Section 2.1 Media Impact and Global Relationships

Media has impacted the leading edge of *global event* reporting in the high-tech age. A few decades ago reporting was through newspapers, radio, and television. Things are different now as we are experiencing a social revolution of people-oriented reporting in *"real time."* This element of intimate knowledge of the *event or story* being reported has dramatically changed the way we all view all of our *global events*. This revolution has intrinsically altered the way *events* are reported. The recent *trends* of people-oriented reporting is on the rise as reporting *events* becomes more personal and more accurate – however more subjective.

Most technologies described as *new media* are digital, frequently having *characteristics* of being: interactive, networkable and manipulated. Some examples are the Internet, websites, computer multimedia, computer games, CD-ROMS, and DVDs. *New media* is generally not described as television programs, films, magazines, books, or paper-based publications - unless they include technologies that facilitate digital interactivity.

There are several ways that *new media* might be described, The *New Media Reader* edited by Wardrip-Fruin and Montfort, defines it by using eight (8) *concise propositions:*

The New Media Versus Cyber Culture - Cyber culture is the social *phenomena* associated with the Internet and network communications (blogs, online multi-player gaming), whereas, the new media is concerned more with cultural objects and paradigms (including digital to analog television, iPhones, etc).

The New Media as Computer Technology Used as a Distribution Platform - new media are the cultural objects that use digital computer technology for distribution. e.g. Internet, web sites, computer multimedia, blu-ray disks etc.

The New Media as Digital Data Controlled by Software - language of the new media is based on the assumption that, all cultural objects that rely on digital representation and computer-based delivery do share common qualities. The new media is reduced to digital data that can be manipulated by software as any other data. Now media operations can create several versions of the same object.

The New Media as the Mix Between Existing Cultural Conventions and the Conventions of Software - new media is understood as the mix between older cultural conventions for data representation, access, and manipulation and newer conventions of data representation, access, and manipulation. The 'old' data are representations of visual reality and human experience, and the *'new'* data are numerical data. The computer is kept out of the key 'creative' decisions, and is delegated to the position of a technician.[27].

The New Media as the Aesthetics that Accompanies the Early Stage of Every New Modern Media and Communication Technology - "*While ideological tropes indeed seem to be reappearing rather regularly, many aesthetic strategies may reappear two or three times...In order for this approach to be truly useful it would be insufficient to simply name the strategies and tropes and to record the moments of their appearance; instead, we would have to develop a much more comprehensive analysis which would correlate the history of technology with social, political, and economical histories or the modern period.*"[28]

The New Media as Faster Execution of Algorithms Previously Executed Manually or through Other Technologies - Computers are a huge speed-up of what were previously manual techniques. "*Dramatically speeding up the execution makes possible previously non-existent representational technique.*" (This also makes possible many new forms of media art such as

---

27   Noah WARDRIP-FRUIN and Nick MONTFORT (eds.), *The New Media Reader*. Cambridge, The MIT Press, 2003.
28   Ibid

interactive multimedia and computer games. *"On one level, a modern digital computer is just a faster calculator; we should not ignore its other identity, that of a cybernetic control device."*[29]

The New Media as the Encoding of Modernist Avant-Garde; The New Media as Metamedia - Manovich declares that the 1920s are more relevant to the new media than any other time period. Metamedia coincides with postmodernism in that they both rework old work rather than create new work. The new media avant-garde is about new ways of accessing and manipulating information (e.g. hypermedia, databases, search engines, etc.). Metamedia is an example of how quantity can change into quality as in the new media technology and manipulation techniques can *"recode modernist aesthetics into a very different postmodern aesthetics."*[30]

The new media as Parallel Articulation of Similar Ideas in Post-WWII Art and Modern Computing - Post-WWII Art or *"combinatorics"* involves creating images by systematically changing a single parameter. This leads to the creation or remarkably similar images and spatial structures. *"This illustrates that algorithms, this essential part of the new media, do not depend on technology, but can be executed by humans."*[31]

Section 2.1.1: Globalization and New Media

The rise of the *new media* has increased and accelerated the communication between people all over the world. It has allowed a wide distribution of views, information and ideas through blogs, websites, pictures, and other user-generated media. Flew stated that as a result of the evolution of the new media technologies, globalization occurs. Globalization is generally stated as *"more than expansion of activities beyond the boundaries of particular nation states."*[32] Globalization has shortened the distance between people all over the world by the electronic communication expressing this great development as the *"death of distance."* These forms of new media *"radically break the connection between physical place and social place, making physical location much less significant for our social relationships."*[33]

---

29 Noah WARDRIP-FRUIN and Nick MONTFORT, *op. cit.*
30 Lev MANOVICH, "New Media from Borges to HTML," in *The New Media Reader*, 2003, Cambridge, The MIT Press, pp. 16–23.
31 Noah WARDRIP-FRUIN and Nick MONTFORT, *op. cit.*
32 Terry FLEW and Sal HUMPHREY, "Games: Technology, Industry, Culture," in *New Media: An Introduction* (second edition), 2005, South Melbourne, Australia, Oxford University Press, pp. 101–114.
33 David CROTEAU and William HOYNES, (2003). *Media Society: Industries, Images and Audiences* (3rd edition). Thousand Oaks, Pine Forge Press, 2003.

However, these changes in the *new media* environment also foster levels of tensions in the concept of what is public information. According to Ingrid Volkmer, *"public sphere"* is defined as a process through which public communication becomes restructured and partly un-embedded from national political and cultural institutions. This trend of the globalized public sphere is not only as a geographical expansion from a nation to worldwide, but also changes the relationship between the public, the media, and state.[34]

Virtual communities online transcend and extend geographical boundaries removing social restrictions. Howard Rheingold describes these globalized societies as self-defined networks that attempt to recreate what we do in real life. *"People in virtual communities use words on screens to exchange pleasantries and argue, engage in intellectual discourse, conduct commerce, make plans, brainstorm, gossip, feud, fall in love, create a little high art and a lot of idle talk."*[35] For Sherry Turkle, *"…making the computer into a second self, finding a soul in the machine, can substitute for human relationships."*[36]

While some of these perspectives suggest technology drives and therefore is a determining factor in the process of globalization, arguments involving technological determinism are generally looked down upon by mainstream media studies.[37] Often academics focus on the processes by which technology is funded, researched, and produced, creating a feedback loop in these the technologies, often influenced and transformed by the users who then feed into the process. Commentators such as Manuel Castells, refers to a soft determinism, contending that technology does not determine society. Nor does society set the course of technological change, since many factors, including individual inventiveness and entrepreneurialism, intervene in the process of technical innovation, social applications and scientific discovery; hence, the outcome depends on complex patterns of interactions. Indeed, technological determinism is probably not a problem, since technology is society and society cannot be understood without its technological tools.[38] This is distinctly different than stating that societal changes are driven by technological development, as highlighted in the theses of Marshall McLuhan.[39]

34 Ingrid VOLKMER, *News in the Global Sphere. A Study of CNN and its Impact on Global Communication.* Luton, UK, University of Luton Press, 1999.

35 Howard RHEINGOLD, The Virtual Community: Homesteading on the Electronic Frontier, Cambridge, The MIT Press, 2000.

36 Sherry TURKLE, "Who am We?" *Wired*, 4.01, January 1996.

37 Raymond WILLIAMS, *Television: Technology and Cultural Form*, London, UK, Routledge, 1974; Meenakshi DURHAM, and Douglas KELLNER, *Media and Cultural Studies: Keyworks (KeyWorks in Cultural Studies)*, Malden, MA and Oxford, UK, Blackwell Publishing, 2001; Martin LISTER, Jon DOVEY, Seth GIDDINGS, Ian GRANT, and Kieran KELLY, *New Media: A Critical Introduction*, London, UK, Routledge, 2003.

38 Manuel CASTELLS, The Rise of the Network Society (The Information Age: Economy, Society and Culture, Volume 1), Hoboken, Wiley-Blackwell, 1996.

39 Marshall MCLUHAN, *The Gutenberg Galaxy: The Making of Typographic Man*, London, UK, Routledge and Kegan Paul, 1962; Marshall MCLUHAN, *Understanding Media: The Extensions of Man*, Toronto, Canada, McGraw-Hill, 1964.

Manovich and Castells both argued that mass media, *"corresponded to the logic of industrial mass society, which values conformity over individuality,"*[40] and therefore, *new media* follows the logic systems of the postindustrial or globalized society whereby *"every citizen can construct her own custom lifestyle and select her ideology from a large number of choices. Rather than pushing the same objects to a mass audience, marketing now tries to target each individual separately."*[41]

Section 2.1.2: Social Change

The social media revolution has proven to be a powerful tool for a wide range of constituencies, from revolutionaries in the Middle East and North Africa to companies in the US and beyond. All types of political, economic, and cultural sectors are particularly interested in social media as a means of creating assemblies where people can learn to articulate, sell, and distribute their adverse views. However, by using the anonymity of the Internet and publicizing as many unfounded allegations as one can craft, the social media revolution can make it look as though there is a contagiousness in seeking this attribute of criticism against a particular country or individual, when in reality, all the claims emerge from a small group of self-interested parties. Despite such controversies, experts such as Andrew T. Stevens, an assistant professor of social media strategy at INSEAD, the international business school in France, have stated: *"If I was a senior executive at a major corporation, I would have this on my radar screen as something to keep an eye on."*[42]

Social movement media has a many-faceted and colorful history that has changed at a rapid rate since the new media became widely used. *The Zapatista Army of National Liberation* of Chiapas, Mexico was the first to make widely recognized and effective use of the new media for communiqués and organizing in 1994.[43] Since then, the new media is used extensively by social movements to communicate, share, educate, organize, promote cultural movements, coalition build, etc. The WTO Ministerial Conference of 1999 protest activity was a landmark in the use of the new media as a tool for social change. The WTO protests used media to organize the original action, to communicate with and educate participants, and as an alternative

---

40   Lev MANOVICH, *op. cit.*; Manuel CASTELLS, *op. cit.*
41   Lev MANOVICH, *op. cit.*, p. 42.
42   Rhea WESSEL, "Activist Investors Turn to Social Media to Enlist Support," *The New York Times DealBook*, March 24, 2011.
43   Chris ATTON, "Reshaping Social Movement Media for a New Millennium," *Social Movement Studies*, 2, 2003, pp. 3–15.

media source.[44] The *"Indy media"* movement developed out of this action and has been a great tool in the democratization of information, which is another widely discussed aspect of the new media movement.[45] Some scholars respond that this democratization as an indication of the creation of a *"radical, socio-technical paradigm to challenge the dominant, neoliberal and technologically determinist model of information and communication technologies."*[46] A less radical view is that people are taking advantage of the Internet to produce a grassroots globalization, one that is (anti-neoliberal) and centered on people rather than the flow of capital.[47]

Many are skeptical of the role of the new media in social movements. Some scholars point out that unequal access exists to the new media creating a hindrance to broad-based movements, sometimes even oppressing some voices within a movement.[48] Others are skeptical about how democratic or useful it really is for social movements, even for those with access.[49] Activists cite that there are many new media components as tools for change that have not been widely discussed as such by academics.

The new media has employed less radical social movements using websites, blogs, and online videos to demonstrate the effectiveness. This approach, the use of high volume blogs allowed numerous views and practices to be more widespread and gain more public attention. An example, the on-going Free Tibet Campaign, which has been seen on numerous websites as well. Another social change seen coming from the new media are trends in fashion and the emergence of subcultures such as: Cyberpunk, Text Speak, and others.

Section 2.1.3: National Security

The *new media* has also become of interest to the global espionage sectors. Accessible electronic database formats can be quickly retrieved and reverse engineered by national governments. A key interest to the espionage community are two sites, *Facebook* and *Twitter*, where individuals divulge

---

44 Thomas Vernon REED, "Will the Revolution be Cybercast?: New Media, the Battle of Seattle, and Global Justice," in *The Art of Protest: Culture and Activism from the Civil Rights Movement to the Streets of Seattle*, 2005, Minneapolis, University of Minnesota Press, pp. 240–285.
45 Douglas KELLNER, "New Technologies, Technocities, and the Prospects for Democratization," in *Technocities*, 1999, London, UK, Sage, pp. 186–204.
46 Paschal PRESTON, Reshaping Communications: Technology, Information and Social Change, London, UK, Sage, 2001.
47 Douglas KELLNER, D. "Globalization and Technopolitics," in *The Future of Revolutions: Rethinking Radical Change in the Age of Globalization*, 2003, New York, Zed Books, pp. 180–194.
48 Herman WASSERMAN, "Is a New Worldwide Web Possible? An Explorative Comparison of the Use of ICTs by Two South African Social Movements," *African Studies Review*, 50, 2007, pp. 109–131.
49 Stephen MARMURA, "A Net Advantage? The Internet, Grassroots Activism and American Middle-Eastern policy," *New Media & Society*, 10, 2008, pp. 247–271.

personal information that can then be sorted, researched and archived for the creation of database files on a growing community of both people of interest and average citizens.

Section 2.1.4: Interactivity and New Media

Interactivity is now a familiar term for a number of the *new media* use options evolving the immediate dissemination of the digitalization of media, media convergence and Internet access points. In 1984, Rice defined *the new media* as communication technologies that enable or facilitate user-to-user interactivity and interactivity between user and information,[50] such as the Internet replaces the *"one-to-many"* approach of traditional mass communication with the possibility of a *"many-to-many"* form of communication. Individuals with the applicable technology can produce online media and include images, text, and sound about whatever they choose.[51] This new media technology shifts the approach of mass communication and radically shapes the way the world is interacting and communicating. Vin Crosbie described new media in, *"What is the new media?"* He saw interpersonal media as *"one to one,"* mass media as *"one to many"* and, finally; the *new media* as *individuation* media or *"many to many."*[52]

In the past, interactivity, assumed a definition related to the conversational dynamics of individuals who are face-to-face. This approach does not allow us to see its presence now in mediated communication forums. It's also viable in the applications of traditional media. In the mid-1990s filmmakers used inexpensive digital cameras to create films. Around the same time, moving image technology capable of being viewed on computer desktops in full motion had developed. Development of new media technology also provided new options for artists to share their work and interact. Other venues of interactivity also include, computer and technological programming, radio and television call –in guest speaker talk shows with listener participation in programs and letters to the editor forums. Interactivity in the new media has benefited everyone because people can express their views in more than one way with the technology.

Interactivity should be considered as a central theme in understanding the *new media*, as different media forms possess different degrees of it. On the other hand, various forms of digitized and converged media are not

---

50 Angela SCHORR, Michael SCHENK, and William CAMPBELL, *Communication Research and Media Science in Europe*, Berlin, Germany, Mouton de Gruyter, 2003.
51 Vin CROSBIE, "What is New Media?" 1998, *Sociology Central*. Retrieved from http://www.sociology.org.uk/as4mm3a. doc
52 David CROTEAU and William HOYNES, *op. cit.*

interactive. Tony Feldman, for example, considers digital satellite television as a new media technology that employs digital compression to dramatically increase the number of television channels that can be delivered, in effect, altering the range of selections of what can be offered by the service providers, viewed from the user's point perspective, it still lacks a more fully interactive dimension. For this example *interactivity* is not a *characteristic* of all the new media technologies, compared to digitization and convergence.[53]

Terry Flew presents the position that *"the global interactive games industry is large and growing, and is at the forefront of many of the most significant innovations in the new media."*[54] Some examples of where interactivity in online computer games are, *World of Warcraft*, *The Sims Online* and *Second Life*. These games, which are product developments of the new media, establish relationships and experience a sense of belonging for users, despite temporal and spatial boundaries. The new media have created virtual realities becoming extensions of the world we live in. These games can be used as an escape or to act out a desired life. The new media changes continuously because it is constantly modified and redefined by the interaction between the creative use of the masses, emerging technology, cultural changes, etc.[55]

Section 2.1.5: Social Media and Data Revolution

The *Social Data Revolution* (SDR) is the shift in human communication patterns towards increased personal information sharing and its related implications, made possible by the rise of social networks in early 2000s. While social networks were used in the early days to privately share photos and private messages, the subsequent trend towards people passively and actively sharing personal information more broadly has resulted in unprecedented amounts of public data. Social data refers to that which individuals create, which is knowingly and voluntarily shared by them. Cost and overhead previously rendered this semi-public form of communication unfeasible, but advances in social networking technology from 2004–2012 have made broader concepts of sharing possible. The types of data users are sharing include geo-location, medical data, dating preferences, open thoughts, interesting news articles, etc. Technology continues to track user behavior with more and more precision, resulting in products and services that can be mass customized to a broad range of users.

---

53 Tony FLEDMAN, *An Introduction to Digital Media*, London, UK, Routledge, 1997.
54 Terry FLEW, *op. cit.*
55 Second Life, http://secondlife.com/whatis/#Be_Creative

Early examples of social data are *Craigslist* and the *"wish lists"* of *Amazon.com*. Both enable users to communicate information to anybody who is looking for it. They differ in their approach to identity. *Craigslist* leverages the power of *anonymity*, while *Amazon.com* leverages the power of *persistent identity*, based on the history of the customer with the firm. The job market is even being shaped by the information people share about themselves on sites like *LinkedIn* and *Facebook*.

Section 2.1.6: Social Authority

One of the key components in successful social media marketing implementation is building *social authority*. Social authority is developed when an individual or organization establishes itself as an expert in a given field or area, thereby becoming an *"influencer"* (per Gladwell) of a tipping point in that field or area. It is through this process of building social authority that social media is very effective. That is why one of the foundational concepts in social media has become that you cannot completely control your message through social media but rather you can simply begin to participate in the conversation in the hopes that you can become a relevant influence in it. These types of *phenomena* were integral and exhibited with the *Facebook* platform use during the recent 2011 *Arab Spring Uprisings*.

However, these types of conversation participation must be cleverly executed because people are, in general, resistant to direct or overt marketing through social media platforms. On the surface, this seems counter-intuitive, but it is the main reason building social authority with credibility is so important. A marketer (or *connector* to use the Gladwell term) generally cannot expect people to be receptive to a marketing message in and of itself. In the Edelman Trust Barometer report in 2008, the majority (58%) of the respondents reported they most trusted company or product information coming from *"people like me,"* inferred to be information from someone they trusted.[56] In the *2010 Trust Report*, the majority switched to 64% preferring their information from industry experts and academics. According to Inc. Technology's Brent Leary, *"This loss of trust, and the accompanying turn towards experts and authorities, seems to be coinciding with the rise of social media and networks."*[57] Thus, using social media as a form of marketing has taken on whole new challenges. As the *2010 Trust Study* indicates, it is most effective if marketing efforts through social media revolve around

---

56 Edelman, *Edelman Trust Barometer 2008*, 2008. Retrieved from http://www.edelman.com/trust/2008/ TrustBarometer08_FINAL.pdf

57 Brent LEARY, "Overemphasis on Brand Building Leads to Mistrust, *Inc.*, March 24, 2010. Retrieved from http:// technology.inc.com/2010/03/22/overemphasis-on-brand-building-leads-to-mistrust/

the *genuine* building of authority.[58] Someone performing a *"marketing"* role within a company must honestly convince people of their *genuine* intentions, knowledge, and expertise in a specific area or industry through providing valuable and accurate information on an ongoing basis without a marketing angle overtly associated. If this can be done, trust with and of the recipient of that information – and that message itself – begins to develop naturally. This person or organization becomes a thought leader and value provider setting themselves up as a trusted advisor instead of marketer. *Top of mind awareness* develops and the consumer naturally begins to gravitate to the products and/or offerings of the authority/influencer.[59]

Of course, there are many ways authority can be created and influenced, for example, the participation in *Wikipedia*, which actually verifies user-generated content and information. Authority also provides valuable content through social networks on platforms such as *Facebook* and *Twitter*; article writing and distribution through sites such as *Ezine* and fact-based answers on *"social question and answer sites"* such as *EHow*. As a result of social media and the direct or indirect influence of social media marketers, consumers are as likely (or more likely) to make buying decisions based on what they read and see in platforms we call *"social"* but only if presented by someone they have come to trust. That is why a purposeful and carefully designed social media strategy has become an integral part of any complete and directed marketing plan (or as we will see in the case studies) but must also be designed using newer authority building techniques.

Social media takes many different forms, including Internet forums, weblogs, social blogs, micro blogging, wiki's, podcasts, photographs or pictures, video, rating and social bookmarking. By applying a set of theories in the field of media research (social presence, media richness) and social processes (self-presentation, self-disclosure) Kaplan and Haenlein created a classification scheme for different social media types in their Business Horizons article published in 2010. According to their findings, there are six different types of social media: collaborative projects, blogs and micro blogs, content communities, social networking sites, virtual game worlds, and virtual communities.[60] Social media technologies include blogs, picture-sharing, blogs, wall-postings, e-mail, instant messaging, music-sharing, crowd-sourcing, and voice over IP, to name a few. Many of these social media services can be integrated via social network aggregation platforms such as:

---

58 Edelman, *Edelman Trust Barometer 2010.* 2010. Retrieved from http://www.edelman.com/trust/2010/
59 Nathan LINNELL, "Social Media Influence on Consumer Behavior," *Search Engine Watch*, May 3, 2010. Retrieved from http://searchenginewatch.com/article/2049190/Social-Media-Influence-on-Consumer-Behavior
60 Andreas KAPLAN and Michael HAENLEIN, "Users of the World, Unite! The Challenges and Opportunities of Social Media," *Business Horizons*, 53, 2010, pp. 59–68.

Facebook, is a social network service launched in February 2004. As of July 2010,[61] *Facebook* had more than 500 million active users. Users typically create a personal profile, add other users as friends, and exchange messages, and it will include automatic notifications when they update their profile. Additionally, users may join common interest user groups organized by workplace, school or college, or other characteristics. The name of the service stems from the colloquial name for the book given to students at the start of the academic year by university administrations in the US with the intention of helping students get to know each other better. *Facebook* allows anyone who declares him or herself to be at least 13 years old to become a registered user of the website.

A January 2009 *Compete.com* study ranked *Facebook* as the most used social network service by worldwide monthly active users followed by *MySpace*.[62] *Entertainment Weekly* published its end-of-the-decade "*best-of*" list, saying, "*How on earth did we stalk our exes, remember our co-workers' birthdays, bug our friends, and play a rousing game of Scrabulous before Facebook?*"[63] *Quantcast* estimates *Facebook* has 135.1 million monthly unique U.S. visitors in October 2010. *Social Media Today*, as of April 2010, estimated that 41.6% of the U.S. population has a *Facebook* account.[64]

Facebook has been met with global event controversies. It has been blocked in several countries intermittently recently including, Egypt, Vietnam, Iran, Uzbekistan, Pakistan, Syria, People's Republic of China and Bangladesh. For example, on the basis of Anti-Islamic and religious discrimination content allowed by *Facebook*, it was also banned in many regions and countries of the world. The privacy of *Facebook* users is also an issue, and the safety of user accounts has been compromised several times. *Facebook* has recently settled a lawsuit regarding claims over source code and intellectual property.[65]

Facebook's role in the *Obama Presidential Campaign* and the American political process was demonstrated in January 2008 when shortly before the New Hampshire primary *Facebook* teamed up with *ABC* and *Saint Anselm College*, allowed users to input live feedback during the January 5th Republican and Democratic debates. Over 1,000,000 users installed the

---

61 Ibid.
62 Andy KAZENIAC, "Social Networks: *Facebook* Takes Over Top Spot, *Twitter* Climbs," *Compete Pulse blog*, February 9, 2009. Retrieved from http://blog.compete.com/2009/02/09/*facebook*-myspace-*twitter*-social-network/
63 Thom GRIER, *et al.*, "The 100 Greatest Movies, TV Shows, Albums, Books, Characters, Scenes, Episodes, Songs, Dresses, Music Videos, and Trends that Entertained Us Over the 10 Years," *Entertainment Weekly*, 1079/1080, December 11, 2009, 74–84.
64 Quantcast, "*Facebook*.com," October 31, 2011. Retrieved from http://www.quantcast.com/*facebook*.com
65 Roy WELLS, "41.6% of the U.S. Population has a *Facebook* account," *Social Media Today*, August 8, 2010. Retrieved from http://socialmediatoday.com/index.php?q=roywells1/158020/416-us-population-has-*facebook*-account

Facebook application "*US politics*" in order to take part, and the application measured users' responses to specific comments made by the debating candidates.[66] This reaction and debate demonstrated that: *Facebook* was an extremely powerful and popular new way to interact and offer opinions. An article written by Michelle Sullivan of *Uwire.com* discussed how the "*Facebook effect*" was influencing youth voting rates, youth support of political candidates, and general involvement by the younger populations in the 2008 election (and 2012?).[67]

Illustrating the global reach of similar trends, in February 2008, a *Facebook* group called "*One Million Voices Against FARC*" organized an event whereby thousands of Colombians marched in protest against the *Revolutionary Armed Forces of Colombia*, FARC.[68] In August 2010, one of North Korea's official government websites, Uriminzokkiri, joined *Facebook*.[69]

Twitter is another social networking website owned and operated by *Twitter* Inc. that offers a service enabling its users to send and read messages called *tweets* via its social networking and micro-blogging platform. Tweets provides for text-based posts of up to 140 characters on the user's profile page. Users may subscribe to other users' tweets—this is known as *following* and subscribers are known as *followers*. Since its launch in July 2006, *Twitter* currently is estimated to have 300 million users as of 2011, generating over 300 million tweets and handling over 1.6 billion search queries per day.[70]

The social platform trend specific to *Twitter*'s popularity was the 2007 *South by Southwest* (SXSW) festival (and actually a tipping point in itself), *Twitter* usage increased from 20,000 tweets per day to 60,000.[71] "*The Twitter people cleverly placed two 60-inch plasma screens in the conference hallways, exclusively streaming Twitter messages,*" remarked *Newsweek*'s Steven Levy. "*Hundreds of conference-goers kept tabs on each other via constant twitters. Panelists and speakers mentioned the service, and the bloggers*

66 Russell GOLDMAN, "*Facebook* Gives Snapshot of Voter Sentiment," *ABC News*, January 5, 2007. Retrieved from http://abcnews.go.com/Politics/story?id=4091460&page=1#.Twrr-2Oonus

67 Michelle SULLIVAN, "'*Facebook* Effect' Mobilizes Youth Vote," *CBS News*, November 3, 2008. Retrieved from http://www.cbsnews.com/stories/2008/11/04/politics/uwire/main4568563.shtml

68 Sibylla BRODZINSKY, "*Facebook* Used to Target Colombia's FARC with Global Rally," *The Christian Science Monitor (Boston)*, February 4, 2008. Retrieved from http://www.csmonitor.com/World/Americas/2008/0204/p04s02-woam.html

69 Laura ROBERTS, "North Korea Joins *Facebook*," *The Daily Telegraph (London)*, August 21, 2010. Retrieved from http://www.telegraph.co.uk/technology/*facebook*/7957222/North-Korea-joins-*Facebook*.html

70 "Your World, More Connected," *Twitter* Blog, August 1, 2011. Retrieved http://blog.*twitter*.com/2011/08/your-world-more-connected.html; *Twitter* Search Team, "The Engineering Behind *Twitter*'s New Search Experience," *Twitter Engineering Blog (blog of Twitter Engineering Division)*, May 31, 2011. Retrieved from http://engineering.*twitter*.com/2011/05/engineering-behind-*twitter*s-new-search.html

71 Nick DOUGLAS, "*Twitter* Blows Up at SXSW Conference," *Gawker*, March 12, 2007. Retrieved from http://gawker.com/243634/*twitter*-blows-up-at-sxsw-conference

in attendance touted it."[72] Other *Twitter* milestones include, the first off-Earth *Twitter* message was posted from the International Space Station by NASA astronaut T. J. Creamer on January 22, 2010.[73] And in late November 2010, an average of a dozen updates per day was posted on the astronauts' communal account, @NASA_Astronauts.[74]

In other developments related to *Twitter, The Wall Street Journal* reported that *Twitter* elicited mixed feelings in the technology-savvy people (who have been their early adopters). However, some users are starting to feel too connected. *Nielsen Online* reported that *Twitter* has a user retention rate of 40%, many people stop using the service after a month, nevertheless, in 2009 *Twitter* won the *"Breakout of the Year" Webby Award.*[75] In February 2009, during a discussion on *National Public Radio's Weekend Edition*, Daniel Schorr stated that: "*Twitter accounts of events lacked rigorous fact-checking and other editorial improvements.*" In response, *Twitter* responded to Schorr with two examples of breaking news stories that played out on *Twitter* and said users wanted first-hand accounts and sometimes debunked stories.[76]

In August 2010, South Korea intermittently blocked content on *Twitter* related to the North Korean government *Twitter* account. The account, setup with user name, @uriminzok, (loosely translated to mean *our people* in Korean), had 4,500 followers in less than one week. Very soon thereafter, on August 19, 2010, South Korea banned the *Twitter* account for broadcasting "*illegal information.*" *BBC*, US and Canada experts claimed that North Korea had invested in "*information technology for more than 20 years*" with knowledge of how to use social networking sites to their power.[77] With only 36 tweets, the *Twitter* account was able to accumulate almost 9,000 followers. To date, the South Korean has banned 65 sites, including this *Twitter* account.[78] *Twitter* is also banned in China (but many Chinese

72 Steven LEVY, "*Twitter*: Is Brevity the Next Big Thing?" *Newsweek*, April 30, 2007. Retrieved from http://www.msnbc.msn.com/id/17888481/site/newsweek/
73 Press release, "Media Advisory M10-012 – NASA Extends the World Wide Web Out into Space," NASA, January 22, 2010. Retrieved from http://www.nasa.gov/home/hqnews/2010/jan/HQ_M10-012_ISS_Web.html
74 Ibid.
75 Staff writer, "13th Annual Webby Special Achievement Award Winners," *The Webby Awards*, n.d. Retrieved from http://www.webbyawards.com/webbys/specialachievement13.php/; Ian PAUL, "Jimmy Fallon Wins Top Webby: And the Winners Are...," *PC World*, May 5, 2009. Retrieved from http://www.pcworld.com/article/164374/jimmy_fallon_wins_top_webby_and_the_winners_are_.html
76 Andy CARVIN, "Welcome to the *Twitterverse*," *National Public Radio*, February 28, 2009. Retrieved from http://www.npr.org/templates/story/story.php?storyId=101265831
77 Clark BOYD, "BBC News – North Korea creates *Twitter* and *YouTube* presence," BBC, August 18, 2010. Retrieved from http://www.bbc.co.uk/news/world-us-canada-11007825
78 Zachary SNIDERMAN, "North Korea's Newly Launched *Twitter* Account Banned by South Korea," *Mashable.com*, August 19, 2010. Retrieved from http://mashable.com/2010/08/19/north-korea-*twitter*-banned/

people use it anyway). Chinese authorities respond to this use very seriously. In 2010, Cheng Jianping was sentenced to one year in a labor camp for a sarcastic post on *Twitter*.[79]

YouTube is a video-sharing website on which users can upload, share, and view videos created by three former *PayPal* employees in February 2005. Unregistered users may watch videos, and registered users may upload an unlimited number of videos. In November 2006, *YouTube, LLC* was bought by *Google Inc.* for $1.65 billion, and now operates as a subsidiary of *Google*. Before the launch of *YouTube* in 2005, there were very few easy methods available for ordinary computer users who wanted to post videos online. With its simple interface, *YouTube* made it possible for anyone with an Internet connection to post a video that a worldwide audience could watch within a few minutes. The wide range of topics covered by *YouTube* has turned video sharing into one of the most important parts of social networking culture.

Section 2.2: Media: Political, Cultural, Economic Relationships

Media has an enormous social networking and micro-blogging social and cultural impact upon society predicated upon the ability to reach a wide audience with a strong and influential message. Marshall McLuhan uses the phrase *"the medium is the message"* as a means of explaining how the distribution of a message can often be more important than content of the message itself.[80] Television broadcasting, for example, has control over the content society watches and the times in which it is viewed. This points out the distinguishing feature of *Traditional Media* that *New Media* has challenged by reinventing the participation habits of the public. The internet creates an opportunity for a heightened level of participation, more diverse political opinions, social and cultural viewpoints. Perhaps suggesting that by allowing consumers to produce information through the internet will result in an overload of information. In a democratic society, all forms of media can serve the electorate about issues regarding government and corporate entities. Some consider the concentration of media ownership to be the greatest threat to democracy. Media can be used for various purposes: For example, it can be used for advocacy, both for business and social concerns. This can include forms of advertising, marketing, propaganda,

79 Andrew JACOBS, "Chinese Woman Imprisoned for *Twitter* Message," *New York Times*, November 18, 2010. Retrieved from http://www.nytimes.com/2010/11/19/world/asia/19beijing.html
80 Marshall MCLUHAN and Fiore QUENTIN, *The Medium is the Message*, Hardwired, San Francisco, 1967, pp. 8–9, 26–41.

public relations, and political communication. Entertainment is another use, traditionally through performances of acting, music, and sports. Third, it can be used for public service announcements.

Another description of Media is *central media* which implies the ability to transmit tacit knowledge. The manipulation of large groups of people through media for the benefit of a particular political party and/or group of people is among its uses. Bias, political or otherwise, towards favoring certain individual outcomes or resolution of an event is inherent. "*This view of central media can be contrasted with lateral media, such as e-mail networks, where messages are all slightly different and spread by a process of lateral diffusion.*"[81]

Section 2.2.1: Relationships: Political

Certain groups tend to promote certain media strategies in an effort to further a political cause. Tipper Gore, the wife of previous US Vice-President Al Gore, was the founder of the *Parents Music Resource Center* and was the main figure in pushing for warning labels on music, even though she does not fit into the conservative demographic. Whereby, demands for the banning of certain songs or the labeling of obscene albums came specifically from conservative political groups in the United States. As such, she argued that such material had simple and identifiable effects on children and thus should be banned/labeled.

Political factions, as perhaps in this instance, sometimes use the media to influence new members into joining and following their groups.

Other political relationships *media strategies* include:

Control

Theorists such as Louis Wirth and Talcott Parsons emphasized the importance of *mass media* as instruments of *social control*. They pointed out, in the 21st century, with the rise of the internet, the two-way relationship between mass media and public opinion is beginning to change with the advent of new technologies such as blogging.[82]

---

81 Ibid.
82 Op cit

Mander's theory is related to Jean Baudrillard's concept of *hyperreality*.[83] Using the 1994 O.J. Simpson trial as an example. The reality reported on in this case was merely the catalyst for the images that defined the trial as a global event, and made the trial more than it should have been. Essentially, hyperreality is where the media is not merely a window onto the world, but is part of the reality described, hence the media's obsession with media-created events.

Marshall McLuhan suggested mass media was increasingly creating a "*global village*." He pointed out, for example, there is evidence that Western media influence (in Asia) is the driving force behind rapid social change: "*It is as if the 1960s and the 1990s were compressed together*."[84] Highlighted by the recent introduction of television to Bhutan. Raising questions of "*cultural imperialism*" — and the de facto imposition through economic and political power and through the media of Western (and in particular US) culture.[85]

Other social scientists have made efforts to integrate the study of the mass media as an instrument of control. For example, in the study of political and economic developments in the Afro-Asian countries, David Lerner, emphasized the general pattern of increase in standard of urbanization, living, literacy and exposure to mass media during the transition from "*traditional to modern society*." Lerner states, "*that while there is a heavy emphasis on the expansion of mass media in developing societies, the penetration of a central authority into the daily consciousness of the mass has to overcome profound resistance*."[86] And, as a recently (2012), the government of Pakistan has even gone so far as to solicit bids ($10 million project) from technology companies to develop a system to cut off the internet service providers with a national-level URL filtering and blocking system. China and other governments, although with very little disclosure have implemented strategies for censorship and sanitizing the internet.

Content

The relation of the mass media to contemporary popular culture is commonly defined in terms of how dissemination occurs from the elite ruling class to the mass. The long-term consequences of significant control and concentration of ownership of the media, lead to accusations of media elite establishing a form of cultural dictatorship, continuing the recent debates about the influence

---

83 Op cit
84 Op cit
85 Op cit
86 David LERNER, The Passing of Traditional Society: Modernizing the Middle East, New York, The Free Press, 1958.

of media barons such as Conrad Black and Rupert Murdoch. Murdoch has had his share of controversy in this regard, for example, *The Observer* in the UK (March 1, 1998) reported the Murdoch-owned *HarperCollins'* refusal to publish Chris Patten's *East and West* because of the former Hong Kong governor's description of the Chinese leadership as "*faceless Stalinists*" possibly being damaging to Murdoch's Chinese broadcasting interests.[87] In this case, the author was able to have the book accepted by another publisher, but this type of content censorship may be a beacon for the future. A related form is that of self-censorship by all level of members in the media in the interests of protecting their own careers.

Influence

It is indeed objectionable when a specific government tries to influence the printers and editors in addition to imposing many media laws. Every type of newspaper and other media forms are somehow under political influence and where there is political influence there will be false and fabricated news that will be published. In most countries, and more recently in developing controls, there are five sectors that are likely to be influenced by governments. These are:

1. *Media*: including newspapers, TV, radio, Internet, and books, etc.

2. *Education*: including schools, colleges and universities, students, and teachers.

3. *Judicial*: namely judges, lawyers, etc.

4. *Medical*: doctors, government or private.

5. *Defense*: members appointed for the protection of a country internally or externally, i.e. police, army, navy, air force, etc.

Agendas

The *agenda-setting process* is perhaps an unavoidable part of large organizations that make up much of the *mass media*. Four main news agencies—*AP, UPI, Reuters* and *Agence-France-Presse*—claim together to provide 90% of the total news output of the world's press, radio and television.[88] Stuart Hall points out that because some of these media agencies produce material that often is good, impartial and serious, they are accorded a high degree of respect and authority. However, in practice the ethics of the

---

87 Nicholas CLEE, "The Bookseller," *Guardian Unlimited*, March 1, 2003, Retrieved from http://www.guardian.co.uk/books/2003/mar/01/featuresreviews.guardianreview30

88 Rajmohan JOSHI, Encyclopedia of Journalism and Mass Communication: Media and Mass Communication, New Delhi, India, Gyan Publishing House, 2006, p. 95.

press and television are closely related to that of the hegemonic establishment providing vital support to the existing order. Independence (e.g. of the *BBC*) is not *"a mere cover, it is central to the way power and ideology are mediated in societies like ours."*[89]

The public is in effect bribed by radio, television, and newspapers into an acceptance of biased, misleading, and status quo agendas. Greg Philo demonstrates this in his 1991 article, *"Seeing is Believing,"* in which he illustrated the 1984 UK miners' strike were strongly correlated with the media presentation of the event, including the perception of the picketing as largely violent when violence was rare, and the use by the public of phrases that had appeared originally in the media.[90]

McCombs and Shaw further demonstrated agenda-setting in a study conducted in Chapel Hill, North Carolina, USA, during the 1968 presidential elections.[91] Here, a representative sample of undecided voters was asked to outline the key issues of the election as they perceived them. Concurrently, the mass media serving these subjects were collected and their content was analyzed. They concluded, there was a definite correlation between the two accounts of predominant issues. *"The evidence in this study that voters tend to share the media's composite definition of what is important strongly suggests an agenda-setting function of the mass media."*[92]

Section 2.2.2: Relationships: Cultural

Fear in the face of huge power is the dictator's traditional tool for keeping people in check, but in fact by cutting off Internet and wireless service, most recently during Egypt's huge protests in January 2011, the president only confirmed that his fear of *Facebook, Twitter*, laptops, and smart phones empowered his opponents and exposed his weakness to the world, resulting in the toppling of his regime.

The new arsenal of social networks also helped to accelerate Tunisia's revolution, driving out the country's ruler of 23 years. It has also helped to ignite an epidemic that spread across the Arab world at breathtaking speed. The Egypt and Tunisia uprisings offer the latest encouragement for another way to look at the situation. While they are away for the world to keep up with friends and make blog posts, they also provide the same technologies

---

89 *Ibid.*, p. 95.
90 *Ibid.*, p. 96.
91 Maxwell MCCOMBS and Donald SHAW, "The Agenda-Setting Function of Mass Media," *Public Opinion Quarterly*, 36, 1972, pp. 176–187.
92 Ibid.

that were hailed as a factor in the Iran Green Revolution. That led to stirring street protest that followed the disputed presidential election. But since the revolt collapsed, the government's role in monitoring the Internet has become a cautionary tale. As we know, the Iranian police eagerly followed the electronic trails left by activists, which led to their making thousands of arrests in the crackdown that followed. The government even posted photos of its hunt for its enemies inviting other Iranians to supply names and addresses. While the Iranian government has become much more adept at using the Internet to go after activists, it has, in fact, become a powerful political and economic force that protects the ayatollah's regime by creating an online surveillance center believed to be behind an army of hackers that it can unleash at any point against opponents. Repressive regimes around the world have fallen behind their opponents in recent years in exploiting the new technologies, and it is not unexpected when aging autocrats face younger, more technically savvy opponents. But in Moscow, to Iran, and to Beijing, governments have begun to climb the steep learning curve and turn the Internet platforms around to their own political devices.

The countertrend debate that the conventional wisdom about the Internet and networking sites inherently tipped the balance of power in favor of democracy is mistaken. A new book, titled *The Net Delusion: The Dark Side of Internet Freedom*, by Eugenie Morozov has made a case in describing instance after instance of strong men finding their way to use the new media to their advantage.[93] After all, the very factors that brought *Facebook* and similar sites' such future success also have incredible appeal for the secret police. A cyber surfing police officer can compile a dossier on a regime opponent without the trouble of street surveillance and telephone tapping required in the pre-net world. In Egypt in January 2011 the president resorted to traditional blunt instrument against the dissident and his crisis. He cut off the communications altogether, yet other countries have shown greater sophistication. While the fact that *Facebook* is a great database for government, now one has to believe that it is doing more good than harm, helping activists for virtual organizations that would never survive if they met face-to-face. But one has to be aware that they are speaking not only to their friends but to their oppressors as well.

Widney Brown, Senior Director of International Law Policy at *Amnesty International*, points out that these technologies are politically neutral. "*There is nothing deterministic about these tools, Gutenberg's press, or fax machine, or Facebook. They can be used to promote human rights or to undermine*

---

93 Eugenie MOROZOV, "The Net Delusion: The Dark Side of Internet Freedom," *PublicAffairs*, 2011.

human rights."[94] In China thousands of commentators are trained and paid to post pro-government comments on the web and steer online opinion away from the criticism of the Communist Party. The same is occurring in Venezuela where President Hugo Chavez first denounced the hostile *Twitter* comments as terrorism and then turned around and created his own *Twitter* feed. Scott Shane, a reporter at *The Times'* Washington bureau, is the author of a book called *Dismantling Utopia: How Information Ended the Soviet Union.*[95] Yet Russia, much the same with the other governments mentioned, has managed to co-opt several prominent news media entrepreneurs with huge website followings that now strongly skew toward pro-Putin and the other anti-Georgia reports that went viral.

In Egypt Mr. Mubarak's government concluded that it was too late for simple monitoring and completely unplugged his country from the Internet altogether in addition to cutting off the country from television. This was a desperate move by an autocrat who had not learned to harness the tools his opponents have embraced.

Section 2.2.3: Relationships Economic

The *media* has a strong economic impact upon society. This is predicated upon the ability to influence a wide audience with a strong economically driven agenda. As previously pointed out, particularly with television broadcasting, and now the internet, each have a large amount of influence over the content society is exposed to and the frequency and times in which it is viewed. This is a distinguishing feature of *New Media* by altering the economic and purchasing habits of societies. The internet creates a (marketing) environment for presenting diverse entrepreneur, government and business viewpoints, and an increased level of consumer economic influence. These forms of media serve economic platforms, generally in favor of government and corporate entities, which typically have the most means to deploy them. Economic relationships and the concentration of media ownership by governments and large corporations are believed by some to be the greatest threat to democracy. *Media* can be used for various economic purposes: For example, it can be used for marketing, both for business and governmental concerns. This can include advertising, propaganda, public relations, and product/service communication. Political and governmental positioning and party economic strategy is another use, traditionally through

---

94 Scott SHANE, "Spotlight Again Falls on Web Tools and Change," *New York Times*, January 29, 2011, para. 12.
95 Scott SHANE, Dismantling Utopia: How Information Ended the Soviet Union, Chicago, Ivan R. Dee, 1994.

use of disseminating and supporting economic driven agendas, not always without bias or (harmful) impacts. Third, it can be used for recent economic and development announcements.

In the same context as political and cultural relationships, economic relationships sometimes involve the manipulation of large societies and countries through these media outlets for the benefit of a particular political party and/or group of people. Bias, economic, political or otherwise, towards favoring certain individual outcomes or resolution of events is inherent.

Section 2.3: Summary

What this chapter proposes to validate about *Tipping Point Theory* and *media* applied in this context is, in fact, the leading edge equation for understanding current and future world events and ultimately, what will drive world societies closer together (or further apart). The questions about the great shift taking place in the world now will prove to be less about culture differences and more about global communication. Newspapers, TV, radio and all the other current newer forms of media technology only reflect and document what is and has occurred. Several more recent, leading edge media and content tracking platforms have emerged as of this version in 2013, with complex algorithms capable of scanning vast databases of information on the internet and synthesize future frequencies of communication trends by topic and subject interest. At this time, these are only beginning to open the possibilities of how these *global event* questions might be quantitatively monitored and analyzed, one site in particular, *www.recordedfuture.com*, establishes some initial ground-breaking benchmarks.

The next chapters will attempt to validate that when looking at specific case studies we can highlight when and why the *tipping points* arrive that result in *global events*, like rare moments in history, and possibly, we may have a unique opportunity to step out from the old protocols of how these events are perceived and influenced and into something new, a moment when one age ends and the future understanding begins.

Part II

Global Events

INTRODUCTION AND METHODOLOGY

urrent history offers a number of interesting examples of *media impact* on *global events* and their relevance to the *Tipping Point Theory*. Each of the following case studies analyzes and contrasts their significant *attributes* and *characteristics* relative to the media and its impact economically, culturally, and politically. The three case studies examined here were selected on the basis of two main criteria:

1. Each exhibited the *sequential progression* of the *attributes* and *characteristics* leading to a defined *moment in time* when circumstances and *events* changed as a result of the influences that lead to a tipping point and,

2. Each opens opportunities to understand *media* relationships to *global events* around us, their impact and the creation of our new reality.

Most *global events* are connected with *media* in their reach and association, however, analysis constraints are not so easily addressed when researching the communication systems being employed and influencing the outcomes. Each of these case studies, however, is characterized by a *media breakthrough* and *development (phenomena)* and its application is salient to each study discussion. And, as in our last case study, *Climate Change*, the *phenomena* is simply not resolved.

Each of the case studies exhibit the *attributes* and *characteristics* central to the question of whether or not *tipping points* exist, can be analyzed, (and possibly controlled), and sometimes do (and do not) give answers to questions that are raised by them. These case studies provide the opportunity to address the questions by elaborating, analyzing and assessing specific examples.

Another element shared by each of the tipping point case studies examined here is that two are within recent memory. While this was not a selection criterion, per se, it shows that when compared, analyzed, and viewed in the context of other similar events in history, it allows their impacts to be observed and assessed in relation to possible future *global events*.

Putting the Tipping Point *attributes* and *characteristics* into an analytical context provides challenges for exhibiting the influences of media over short and long period of times. Even with the evidence of the particular *characteristic*, some of these influences may be difficult to identify. However, for each *attribute* and *characteristics* identified in each of the case studies, a bracketed timescale is utilized on the order of years as a finite duration for the analysis, and this timescale will then enable us to qualitatively measure and exhibit, on a relative scale, the media impact in terms of the tipping point *attributes* and *characteristics* as researched and identified for each global event.

The three case studies highlight that there are *people, organizations, media, and events* inherently critical to development and potential for a tipping point to occur, and with basis for influence from the media sectors. Progressively, each case study expands the duration timescale *"window"* of the *attribute* and *characteristic* analysis, from the relatively short timeframe of the *Obama Presidential Campaign* of one year (2008 plus), to the four-year (plus) duration of the *International Financial Crisis of 2007–2010*, and the evolving thirty plus years duration focus on *Climate Change* (2000–2030 and beyond).

The purpose is three-fold:

1. Contrasting and comparing the *attributes* and *characteristics* from a very finite timescale to a broader scope to identify similarities and differences,

2. Identifying fundamental philosophies, trends, and media delivery systems and technology that may be prevalent, and

3. Providing for the three case studies to be measured qualitatively relative to the identified *characteristic* importance and influence on the tipping point *attribute* (one dramatic moment).

The following should be noted regarding the limits to quantity of *characteristics* identified for the purposes of sizing the research scope and magnitude accordingly for this version.

First, the research continues to evolve and grow and, perhaps inadvertently, the content on relevant issues was unjustly included given recent discoveries that have come to bear on the potential application of the theory. For this merely reinforces the dynamics of how media is in many ways captured by the changing flux of these phenomena as well.

And second, the research and data is limited given the breadth of information that exists and the fundamental complexity of this application. This has indeed confirmed the admiration for the scientists, scholars, media sources, authors and world leaders whose work and fine efforts that proceeded and was deemed salient to the discussion. In as much as was practical and applicable in the scope of this work, the attempt was to assimilate these perspectives and findings herein and duly note and credit their origins and context with the resources, fact-checking and formatting approaches and graphic and text tools available. If ever there was a shortcoming in this approach, was the focus on the *theory* analysis, with a genuine objective develop and answer questions as they arose and ultimately exhibit the findings and conclusions for future events.

In this context (and current research formatting protocol), invariably, some rules were adjusted regarding the research to move the project forward and support the arguments and not be constrained by these issues. In the following sections, we discuss the methodology and illustrate its application to each of the case studies.

Methodology

What is developed herein is a *qualitative ratio analysis framework* for graphically exhibiting the relative *scalable media impact* with each timescale and for each *attribute* and *characteristic* identified. Each case study chapter identifies a broad range of *characteristics* that correlated to a *media* influence on the evolution of the tipping point. As a principal domain of this application we show how one might approach and provide a coherent representation of the tipping point *attributes, characteristics,* and relationships to *global events* as presented over several different durations of time, subject to some qualitative interpretations but nevertheless accurate enough to permit qualitative analysis.

The *qualitative ratio analysis framework* provides the opportunity to analyze the different *attributes* and *characteristics* and compare their performance measured by the qualitative scale. In the same way that financial experts analyze annual reports, we look at the *attributes* in succession (sequentially), i.e. time duration and use this tool to identify patterns relative to the tipping point *attributes* and *characteristics*. Similarly, this *framework* allows each case study to be interpreted, analyzed, and compared.

Case studies used in this manner are sometimes used to describe *phenomena*. For example, findings about mental health in the longitudinal case studies conducted by Vaillant.[96] The adaptation of this framework as a viable research tool has emerged, in part, as a convenient and meaningful technique to capture a time-framed picture of an aggregate of values that can be construed as a unit or collective-characteristics and performance. While methodological disagreements among practitioners of this approach might continue, nevertheless as of this writing, there is an obligation to reveal how the research was conducted and how collected evidence was handled and interpreted.

In order to be convinced that this approach has merit, we had to develop a relationship between *argument* and *evidence* applying a best practice approach to make this determination; in short, to increase the believability of my findings whereby to identify early in the analysis a clear statement of the conceptual relationship of the *attributes* and *characteristics* and readily determine how this *qualitative ratio analysis framework* translated the researchable questions or issue, or series of questions or issues. In addition, these specifically relate to the nature of the case study undertaken. The longer the time frame over which the case study is conducted, the more difficult it was to ensure accurate representation, hence the need for an applicable framework adaptable to all three case studies for analysis and comparison purposes.

There are four (4) *major issues* relating to the ability to convince and generalize the application of the qualitative ratio analysis framework:

First, context(s) to be represented in a conceptually clear manner so that the reader will find them *convincing*.

Second, findings to be provided from collected information that has been *verified*.

Third, evidence to be provided that the case has been conducted in a manner that is consistent with the principles of *trustworthiness* - in particular the type and extent of triangulation and the presence of an audit trail should have been documented- as described by Lincoln and Guba (1985)[97] or, stated differently, that the criteria for internal and external validity as explicated by Yin (1994).[98]

---

96 George E. VAILLANT, *Adaptations to Life*, Boston, MA, Little Brown, 1977.
97 Yvonne S. LINCOLN and Egon G. GUBA, *Naturalistic Inquiry*, Beverly Hills, CA, Sage, 1985.
98 Robert K. YIN, *Case Study Research: Design and Methods* (2nd ed.) Newbury Park, CA, Sage, 1994.

Fourth, analysis and comparisons to be *believable*. Due to the considerable variation of each of the case studies and general nature of the qualitative research, I felt I had an obligation to exhibit how evidence is interpreted as well as identifying my own viewpoints as opposed to just researcher's footnotes.

Researchers have a number of choices they can make when interpreting evidence to help the reader follow their perspective, such as using computer-based interpretations like Atlas/ti (Muhr, 1997),[99] using constant comparison (Strauss & Corbin, 1990),[100] or a procedure that emphasizes contextual quotations and head-notes (Kvale, 1996; Strauss, 1987).[101] The selected procedure will, of course, vary as a function of perspective and is not limited to the illustrations given.

This illustrative, graphical, and visual approach provides important information gathered and analyzed during the research in the context of the case studies. And through a qualitative (numerical) ratio framework to judge the evidential basis of each case study (Bachor, 2000; Davis & Bachor, 1999),[102] this suggests that a ratio can be computed. This ratio is the *value* and *element point* raised within a *characteristic* in each of the three *attributes* identified within each case study. The research was for a similar set of that was persistent in all three of the case studies. In particular, in the *Obama Presidential Campaign* study, it was observed that the *media lag time* was particularly short in relation to influence and impact, particularly with the Internet. This reinforced the fact that it had a more *"real time"* definition, similar patterns of influential *characteristics* that were transferred quickly via the media.

Moreover, it is the direct influence between *media* and *global events* that exhibits the *sequential tipping point attributes* that evolve and compete in context with other *events* and media contributions that propagate and are diffused through media. Of note is that very few media events evolve and compete for attention within a relatively large set of broader topics that are evident in the media environment. Identification of these case study *characteristics* has proved difficult, and some *characteristics* presented themselves continuously, crossing over *attribute* timeframes, appearing to grow and then diminish, while another *characteristic* during the same

---

99 Thomas MUHR, *Atlas/ti: The Knowledge Workbench, Version 4.1*, Berlin, Scientific SoftwareDevelopment, 1997.
100 Anselm L. STRAUSS and Juliet M. CORBIN, *Basics of Qualitative Research: Grounded Theory Procedures and Techniques*, Newbury Park, CA, Sage, 1990.
101 Steiner KVALE, *Interviews. An Introduction to Qualitative Research Interviewing*, Thousand Oaks, CA, Sage, 1996; Anselm L. STRAUSS, *Qualitative Analysis for Social Scientists*, Cambridge, UK, Cambridge University Press, 1987.
102 Daniel G. BACHOR, "Rethinking Case Study Research Methodology," paper presented at the Special Education National Research Forum, Helsinki, May 2000; T M. DAVIS and Daniel G. BACHOR, "Case Studies as a Research Tool in Evaluating Student Achievement," paper presented at the Canadian Society for Studies in Education Conference, Sherbrooke, Canada, June 1999.

attribute timeframe may have presented itself without significant shifts in the media attention or impact. As a result, the dynamics of this analysis could be the subject of intense interest to researchers in media, the political process, and related global issues, with the focus here being mainly qualitative, it may indeed overlook some scientific techniques for undertaking the quantitative analysis of the question as a whole.

Utilizing these tipping point *attributes* and *characteristics* and constructing the *qualitative ratio analysis framework* as defined above, it is possible to effectively exhibit these ratios for these *characteristics* for each of the tipping point *attributes* and see their relative context. Again, an important point to note at the outset is that the total number of *characteristics* of each *attribute* is by no means inclusive and would require additional study and analysis. Rather, the intention here has been to highlight those *characteristics* that are salient to the analysis of each *attribute*. The approach was to identify a minimum set of *characteristics* for each tipping point *attribute* that can be analyzed and exhibited for the three case studies that exhibit a relationship to our question of *media impact* on *global events*.

Another question that has grown out of this research is:

When does the media impact occur, before or after the global event?

One common assertion is the *event* occurs, and then the *media* propagates information about it after. However, that may have been case with *old media* types, but certainly this notion has lost its credibility given the *real-time* influences of current and future media technology. Another important *characteristic* question that needs to be posed is:

Whether the people, organizations, media and global events can be truly labeled in such a manner? Are these the only characteristics to consider?

To address this question and provide a filter that works well in practice, the research only used *characteristics* that were readily identified, via *Google* or related search engines, books and journals assembled during this project, libraries and archives visited, and firsthand personal observations and accounts. Otherwise there would be such a multitude of *characteristics* to identify and analyze that each attribute "*dataset*" would be overwhelming.

The *qualitative ratio analysis framework* for analyzing the distinctive *characteristics* for the tipping point *attribute* timeframe is developed with the intention of presenting *scalable* (numerical) values for identifying and comparing them. These efforts are some of the first *qualitative ratio analysis* of how the *Tipping Point Theory* can be applied to *media impact* on *global events* and the dynamics of how information is propagated between media and the public. It also points out, in particular, the existence, or lack thereof, of a perception of the "*lag time*" between the *peak of attention* in the media and the *respective influence* on the global event.

Using this tipping point methodology in these case studies certainly provides one interpretation and potentially *one solution for our third and most pressing case study: Climate Change*. Media may hold some very real options for how we as a civilization go about the business of the future of the planet. This approach opens up the opportunity to pursue additional questions that before had been effectively impossible to conceptualize. For example;

How can we characterize the dynamics of media impact on key global events?

How does information change and influence the evolution of the tipping point baseline as it propagates?

Might it be possible to use this tipping point model over longer periods of time, in a way that is essential to discovering the core solution to *climate change?*

One could combine the approaches here into several multidiscipline approaches towards addressing scientific, political, cultural and financial issues associated with this phenomenon. And more generally, it appears useful to further understand the media impact on the different tipping point attributes and characteristics as exhibited during *global events*.

Extracting from the research, we analyzed the case studies with the goal of understanding the *attributes, characteristics* and relationships between them as they unfold. Looking at the research questions that were posed earlier, one can see how the *attributes* took place in relative groupings of *people, organizations, media, and events*, and the net results. This identifies and/or quantifies the relative impact among them. This study of the relationships evolves and becomes more important as we analyze their impact on the outcome. In any case, with the third case study, *Climate Change*, the overall research question is trying to point out that by analyzing these *attributes* it might influence and predict and/or help avoid a potential *global tipping point* from occurring.

An explanation of the *notation* for analyzing the relationships and *attributes* is in order, as *qualitative ratio analysis framework* identifies *people, organizations, media impact,* and *events* that occurred during the case study and influenced the ultimate outcome. One way to measure and determine the *one dramatic moment* of each is to quantify the relative impact of the various issues and in context with others of scalable importance. The aim is to provide a sense of *qualitative influence* and *pneumatic control* as to how these various *attributes* led to the ultimate outcome.

With this tipping point analysis three attributes are identified in the *Qualitative Ratio Analysis Framework*:

- *Contagiousness (C)*

- *Stickiness (S), and,*

- *One Dramatic Point (TP)*

The *notations* indicate either a positive attribute relationship, i.e., $+C$, prevalent with the originating issue, or a $-C$, indicating a negative relationship and so on. The tables and graphics that follow summarize the qualitative research on the various *attributes* recalling these relative values and attempt to identify important relationships in relation to the *events* and *media* influence. The important point is that the relative relationships between the *attributes* are chosen based upon the research of the subject matter, evaluation, and in some cases, researched opinion.

For the purposes of this *tipping point analysis*, the values for the relationships between the *attributes* and the *characteristics* range between values summarized between 1 to 10, or, a *raw score*, with 1 being a very weak indicator and 10 being a very strong indicator. And as a way of understanding the approach and conveying our information effectively, what is also shown are graphs illustrating a *smooth score*, which is helpful when depicting our analysis using line graphs, especially over time. With the large number of characteristics in each of the case studies such as ours, whereby values that zigzag up and down a lot, the key is to try to separate out the meaningless or temporary fluctuations from the underlying, long-run changes, transitions and trends that occur central to the characteristics. Essentially, the *smooth score* substitutes for the *raw score* a new number, which is the average of the characteristic immediately before and after it. The intent is to present the data by reducing the chart information to the essentials.

The logic of our tipping point analysis *qualitative ratio analysis framework* is that these rules govern the identification of key areas of the case study research and provide a means of control over what *characteristics* and issues were included as part of each case study. For example, the influence of the media and the messages delivered by and with the current media technology that existed in 2008, providing the tipping point for presidential hopeful Barack Obama over his rival John McCain during the presidential election is fundamental to that analysis.

Easily, one can interpret these *attributes* and *characteristics* from many perspectives; however, the intent of our tipping point analysis is to define where and when the *one dramatic moment* occurred during each case study and compare this to the other two case studies for a similar analysis. This *qualitative ratio analysis framework* for analyzing these *attributes* is simply a look at the *sequence* of *events* in a chronological manner to observe the way they each developed and how, over the duration of the case study life, *media* influenced each *global event*.

Author's Note: (2013 Version): A complete and comprehensive detail of all the research for each of the three case studies highlighted here is published in: *Media Tipping Points*: Dr. Philip Gordon, PhD, Blue Matrix Publications (2012).

CHAPTER 3: OBAMA PRESIDENTIAL CAMPAIGN

Section 3.1: Introduction

I magine you are born in 1961, and one day, 47 years later you run for the presidential office of the most powerful country on the planet. During your lifetime, major media technological tools become at your disposal to deploy (and guarantee) that you will win the US presidential race. This is why Barack Obama became the 44th president of the USA. The first case study examined the 2007–2008 *Obama Presidential Campaign*. The first case study illustrates that prior to the start of the campaign, during the campaign, and on the election day itself, *global media* echoed the *contagiousness* and the *stickiness* for his victorious campaign, which then lead to the *one dramatic moment*, (tipping point), that dealt a devastating loss for the ill-conceived campaign of Senator John McCain and Governor Sarah Palin.

This case study analyzed the presidential journey of Barack Obama that occurred during the presidential elections of 2008 in the United States. Barack Obama's revolutionary presidential campaign resulted in a culmination of never before experienced voter turnouts and added emphasis to the International Financial Crisis (2007–2010), which was analyzed in *Case Study Two*.

This study explored the Tipping Point *attributes* and *characteristics* of the presidential campaign of Barack Obama. The Tipping Point attributes include *Contagiousness, Stickiness*, and the *One Dramatic Moment*, as each unfolded during the presidential race between Barack Obama and John McCain. These *attributes* and *characteristics* were influenced by the various *media* developments and technological advancements outlined in the earlier chapter that helped to shape the message of the campaigns, how they were delivered, how they were reported on, which events had significance, and ultimately, the voter response.

The goal of this research was to understand the relationships between and among the *attributes* and key *characteristics* individually and collectively. It is these relationships that shaped the campaign and its eventual outcome. The evidence of this case study suggests that the outcome didn't happen in a vacuum, it was the coalescence of the *attributes* and *characteristics*, as a media and social phenomenon led to the culmination of Barack Obama being elected as president of the United States.

Two research questions were paramount in development of this study:

1. *How did analyzing the media delivery systems and technology used during the campaign, exhibit their influence on the reaction of the American voter?*

2. *What relationships between and among the various media delivery systems resulted in this outcome for this historic presidential campaign?*

Answering these questions as part of the case study takes the analysis to a higher-level.

Integral to this case study was a review of the actual campaign personalities and individuals, organizations, media agendas, technology structures and significant events that occurred and existed in 2008 to understand the conditions that proved dynamic for the evolution of the campaign. Also included are brief reviews of the historical state-of-the-art media in order to understand its contribution. (See Authors Note *(p.53)*: Full research documented and published: *Media Tipping Points*, (2012)

This research was formatted to develop this *qualitative ratio framework* to analyze the various *attributes* and *characteristics* of the *Tipping Point Theory*. Part three's goal is to analyze and compare the complexity and the application of this theory. With these research questions in hand, this *framework* provides a methodology that can be applied to other *global media events* and similar *tipping point phenomena* in the other two case studies. The discussion that follows is based upon qualitative research from primary and secondary sources, and the bibliography and further references related to this case study appear at the end. (See Authors Note *(p.53)*: Full research documented and published: *Media Tipping Points*, (2012)

In this case study, we evaluated the Obama Campaign in the context of the US financial sector economic collapse, highlight many of the developments in the media, and analyze the campaign themselves. We examined how many skilled professionals orchestrated these campaigns and masterminded media strategies along the way for each of the two candidates.

Knowing when the *tipping point* occurred for the Obama Campaign is crucial in determining the *media impact* in effecting the 2008 election outcome. The research focused on the *contagiousness* and *stickiness* of the communications that were disseminated by the campaigns and how the *media* treated it and, ultimately, how the voters responded. It explored McCain's age and Obama's race. In addition, it evaluated how the outcome was tied to the unpopular incumbent George W. Bush (and how the McCain team tried to disassociate). The *contagiousness* and *stickiness characteristics* of the vice presidential candidates were also explored.

As part of this case study research, we examined the early developments and key issues in the summer of 2008, which candidate had the better energy policy, the competing agendas over the alternative tax policies, and finally, right up until the election itself, how the plans for saving the economy became so important. In the end, we proposed an account of how the complexity of the *attributes* and *characteristics*, the strategies of the campaigns themselves, and the media technology were at work leading to the tipping point in the campaign and the ultimate result.

The case study was divided by the three (3) *attributes*:

Contagiousness: The effects of the *trends* that resulted in an *epidemic* of forces in the messages that were used by the various sectors of media during the campaign. In doing so, we explored how the candidates' messages about themselves and their opponents played a large part in this *attribute* development.

Stickiness: Analyzing the shifts in *momentum*, some of which seemingly started *small*, ultimately leading to a *larger* impact during the campaign. These *stickiness characteristics*, under the right circumstances, made the possibility of Obama's election a reality.

One Dramatic Moment: When the *epidemic* resulted in a fundamental and irreversible/ irresistible? change in direction for the campaign. The economic collapse led to the tipping point and the ultimate win for the Obama campaign. Understanding this final *attribute* makes sense of the first two and perhaps permits the greatest insight into why these *events* happen the way they do.

This approach was subdivided further into the four (4) specific sets of *characteristics* for each *attribute*: *people, organizations, media* and *events*.

The approach for this case study was to analyze the media delivery systems and technology available that were being used during the campaign and influenced the tipping point *attributes*.

Important to the analysis, is the *contagiousness* and *stickiness characteristics* resulting in voter influence for the candidate, does the analysis indicate a direct relationship between the "*real time*" characteristic referenced and Obama's victory?

Extracting from the research (See Authors Note: Full research documented and published: *Media Tipping Points*, (2012), we analyzed the *attributes* with the goal of identifying the *characteristics* and understanding the relationships between them as the campaign unfolded.

With the research question above and the analysis methodology, what follows are the tipping point *qualitative ratios* and *graphic summary* of how the *attributes* and *characteristics* were influential relative to *people, organizations, media* and *events*. It *identifies* and *quantifies* the relative impact and relationships among them.

This analysis of the *attribute* relationships evolves and become more important in context of the impact on the outcome. The same series of rules governed the research and provided a means of control over what *characteristics/issues* were included as part of this case study. The influence of the *media* and the messages delivered by and with the technology that led to the tipping point for the Obama Presidential Campaign was presented. The influence of the *media* and the messages delivered by and with the technology provided for the *tipping point*.

Clearly, one can interpret these *characteristics/issues* and *attributes* from many perspectives; however, the intent of our *tipping point analysis* is to define where and when the one dramatic event occurred during the Obama Presidential Campaign and then compare these in *Chapter Six* to the other two case studies and *global events*. This method for analyzing the *attributes* is simply to look at the *sequence of events* in a chronological manner and observe the way the campaign developed and how the various *media* sectors played their role. The relationships between the *characteristics* and various *media* delivery systems led to a historical *tipping point* in the 2008 presidential campaign.

The following *qualitative ratios* and *attributes* reveal an actual correlation between media and the outcome of the campaign:

Section 3.2: Contagiousness Attribute (C)

| Date: | 05-Sep-11 | | Start Time Frame: | October 1, 2007 |
| Revised: | | | End Time Frame: | February 1, 2008 |

Obama Presidential Campaign: Contagiousness Attribute

Qualitative ratio scale	McCain Raw score	McCain Smooth score	Obama Raw score	Obama Smooth score
Neutral 0				
Characteristic				
People	0	0	0	0
George Bush	-2	-1	2	2
Economy	-3	-2	3	3
Iraq war	-3	-1	3	3
Public opinion	0	0	4	3
John McCain	0	0	0	2
Veteran/proven	3	1	-2	0
Temper/out of touch/too old	-2	0	2	-1
Barack Obama		0		-1
Risky/not ready to lead	3	1	-2	-2
Rev. Wright	3	2	-3	-3
Racial issue	2	2	-4	-3
Underworld connections	2	1	-5	-2
Organizations	0	0	0	0
American voter/frustration	-5	-3	5	3
Opinions/war/economy	-6	-5	5	4
Previous elections	-6	-3	4	4
Republican/conservative	2	-2	-2	3
Democrats/liberals	-2	-1	2	2
Bush	-3	-2	3	3
Iraq	-5	-4	5	4
Taxes	-5	-3	5	5
Economy/who could solve?	0	-2	0	4
McCain	-2	-2	4	4
Obama	-3	-2	4	4
"Chicago machine"	0	-1	3	3
Media	0	0	0	0
"Real Time"	-5	0	5	3
Radio	3	2	3	2
Internet news	2	2	3	2
Google stories	0	1	2	2
TV news	1	1	3	2
Economy	-2	0	3	3
Ad campaigns	1	1	4	3
Age	-1	0	2	2
Two young	1	0	0	1
Trustworthy	2	1	-1	0
Internet "viral"	-1	0	1	0
US Citizen	1	0	0	0
Events	notes: do not add*	0	0	0	0
	space *				
	space *				
	space *				
Economy	note position on chart	-3	-2	3	2
	space *				
	space *				
DOW	note postion on chart	-4	-3	4	3
	space *				
	space *				
Wall Street	note position on chart	-5	-3	5	3
	space *				
	space *	0	0	0	0

Table 3.2: Qualitative Ratios: Obama Campaign:
Contagiousness Attribute - Raw and Smooth Scores

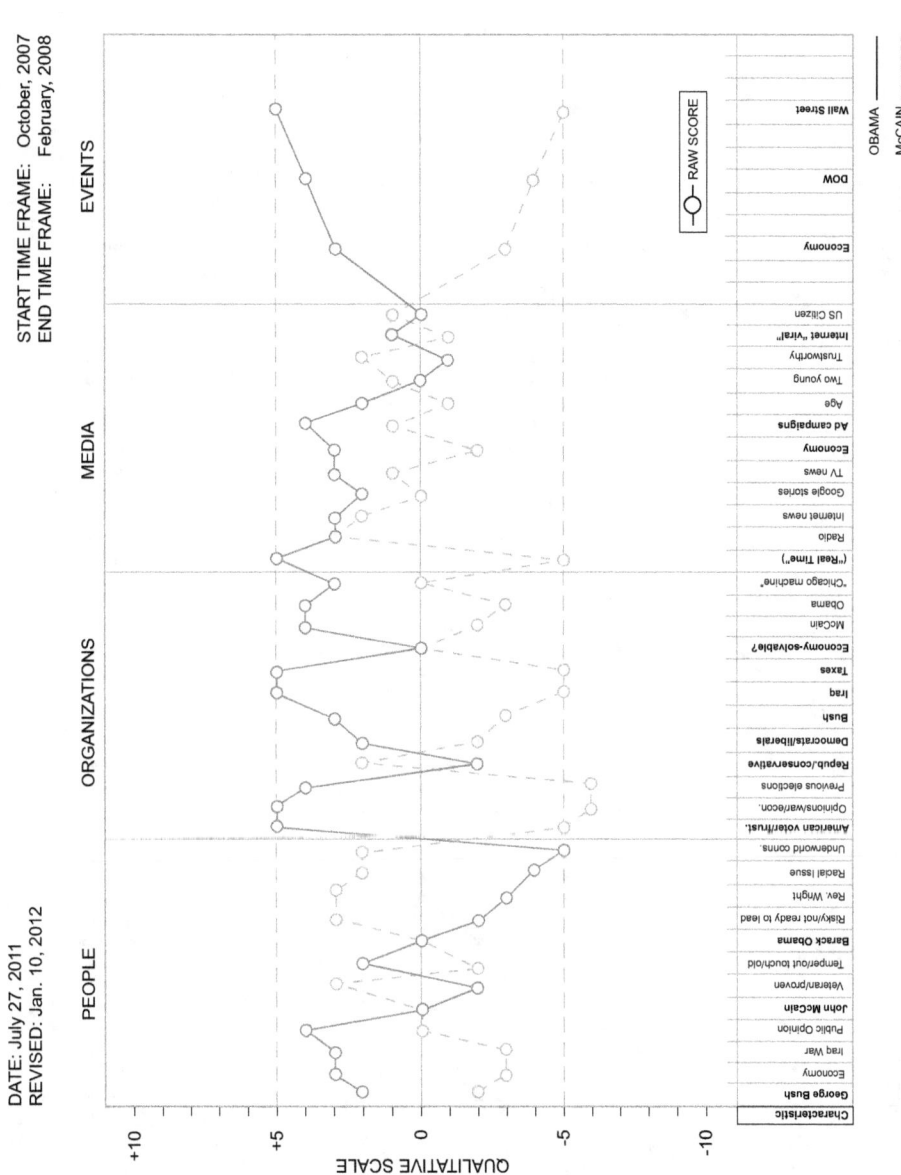

Chart 3.2 (r): Obama Campaign: Contagiousness Attribute - Raw Scores

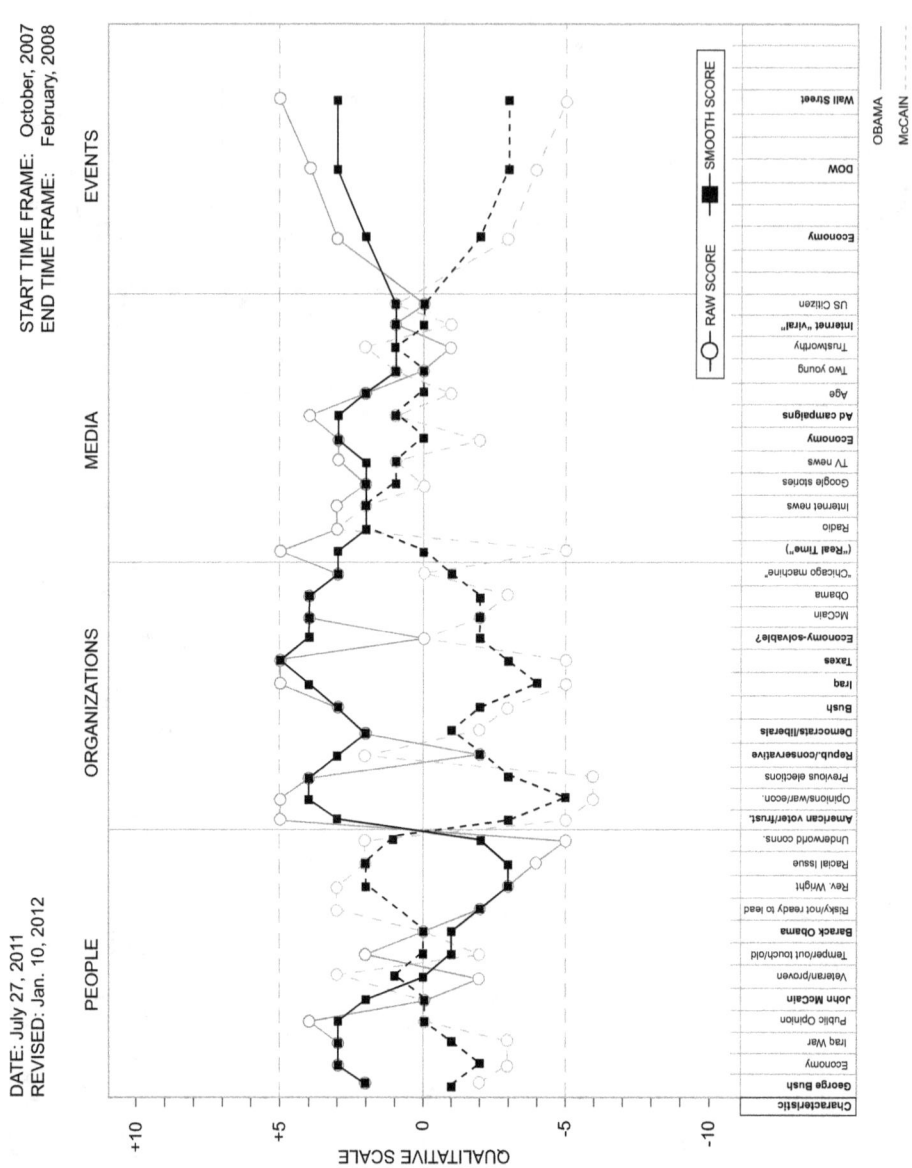

Chart 3.2 (rs): Obama Campaign: Contagiousness Attribute -
Raw and Smooth Scores

Section 3.3: Stickiness Attribute (S)

Date: 05-Sep-11			Start Time Frame: February 1, 2008	
Revised:			End Time Frame: June 1, 2008	

Obama Presidential Campaign: Stickiness Attribute

Qualitative ratio scale	McCain		Obama	
	Raw score	**Smooth score**	**Raw score**	**Smooth score**
Characteristic				
People	0	0	0	0
Joe Biden	-4	-2	6	3
Voting record	-2	-2	2	2
Sarah Palen	-3	-3	1	1
Experience	-5	-3	2	2
Gender	-3	-3	1	1
"Trooper gate"	-5	-3	3	2
McCain		-3		1
"Heartbeat away"/age	-5	-3	4	2
Competence	3	0	0	1
Obama		0		0
Budget experience	4	1	0	2
Media exposure	-3	0	4	3
Organizations	0	0	0	0
Democratic	0	0	0	0
Convention	-4	-2	5	3
"Change versus same"	-5	-3	6	4
Biden speech	-6	-3	6	5
Obama speech	-6	-2	6	3
Republican		0		1
Convention	2	1	0	0
Palen speech	3	2	0	0
"Maverick appeal"	5	3	-1	0
Distance Bush	-1	1	0	0
Media	0	0	0	0
"Real Time"	-5	-3	5	3
Economy capability	-6	-4	3	3
Wall Street/perception fault	-7	-5	6	5
Bush link	-8	-4	7	6
TV debates	-2	-3	7	6
Tax increases	3	-2	3	5
VP candidates/speeches	4	-1	5	5
Speeches/ad campaigns	1	-1	7	6
Shared values	-2	-2	5	5
Technology	-5	-3	7	6
Frequency	-2	-2	8	7
Agenda	2	-1	6	5
Events	0	0	0	0
Energy/gas	3	1	5	3
Tax issues	3	1	4	4
DOW drop	-3	0	6	5
Wall Street	-4	-1	7	6
Offshore drilling	3	0	-3	4
Obama lead?	-5	-2	5	4
National security	3	0	0	3
Economy	-7	-3	5	4
Hurricane	0	-2	2	3
Iraq war	-4	-3	3	3
Afghanistan	-5	-4	4	4
Public anxiety	-7	-5	7	5

Table 3.3: Qualitative Ratios: Obama Campaign:
Stickiness Attribute - Raw and Smooth Scores

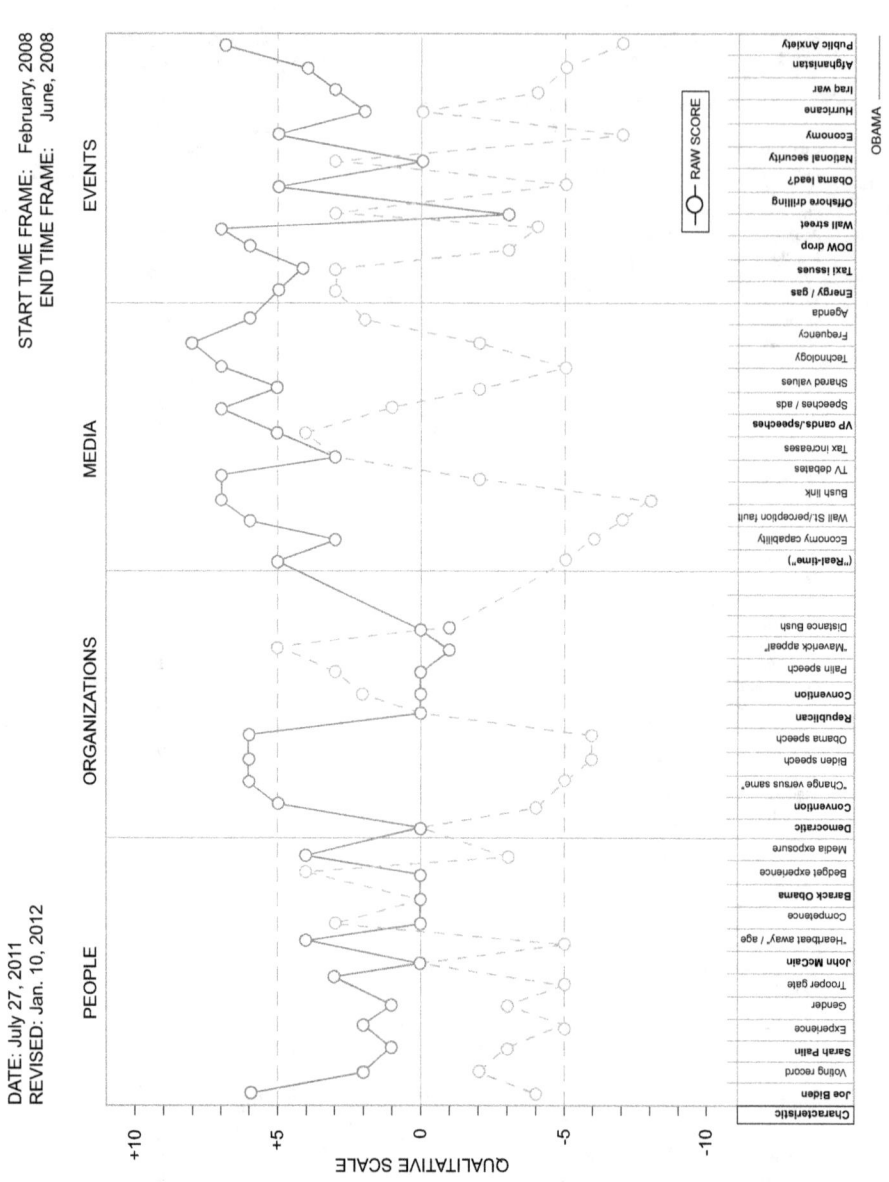

Chart 3.3 (r): Obama Campaign: Stickiness Attribute - Raw Scores

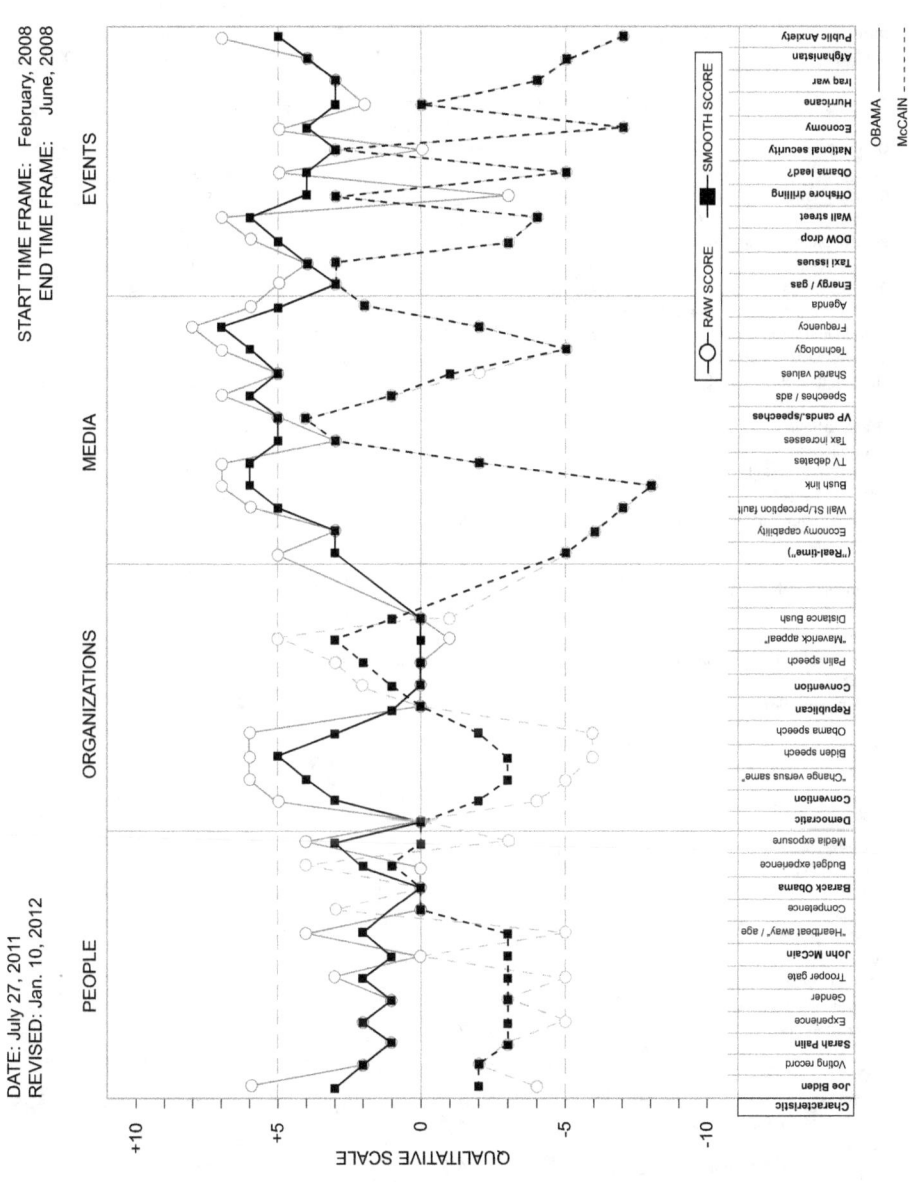

Chart 3.3 (rs): Obama Campaign: Stickiness Attribute -
Raw and Smooth Scores

Section 3.4: One Dramatic Moment: The Tipping Point (TP)

Date:	05-Sep-11		Start Time Frame:	June 1, 2008
Revised:	24-Sep-11		End Time Frame:	November 6, 2008

Obama Presidential Campaign: One Dramatic Moment Attribute

	McCain		Obama	
Qualitative	Raw	Smooth	Raw	Smooth
ratio scale	score	score	score	score
Characteristic				
People	0	0	0	0
Public opinion polls	-5	-3	5	3
John McCain	-6	-4	6	4
Barack Obama	-5	-5	7	6
Unemployment: September 5, 2008	-6	-6	8	7
"Fundamentals of the economy" September 15, 200	-9	-7	10	9
Debates: October 15, 2008	-6	-6	7	8
Election day: November 5, 2008	0	0	9	9
Organizations				
Republicans	0	0	0	0
"Economic freefall"	-7	-3	8	4
"Out of touch"	-6	-5	6	6
"Fundamentals of the economy are strong"	-9	-6	10	8
Democrats	0	-3	0	7
Link Bush	-5	-5	5	8
"Fundamentals"	-9	-3	10	9
Strength of campaign	0	0	7	7
"Spread the wealth"	3	1	-3	3
"Joe the plumber" October 12, 2008	5	3	-4	2
Debates 15-Oct-08	-1	1	6	3
Banking situation	-3	-1	3	0
Media	0	0	0	0
"Real-time"	-5	-1	7	4
Radio	2	0	5	5
Internet news	2	1	6	6
TV news	3	2	6	6
Ad campaigns	3	2	8	7
Internet "viral"	-3	0	7	7
E-mails	-7	-5	7	7
Text messaging	-6	-6	7	7
Videos	-6	-6	5	6
Cell phones	-3	-5	5	7
Digital technology/innovation	-5	-5	9	8
Micro-targeting	-7	-5	8	7
Events	0	0	0	0
Economy	-5	-3	5	3
DOW October 15, 2008 735 points	-5	-5	5	4
Wall Street	-7	-7	5	5
"Fundamentals": September 15, 2008	-9	-9	10	7
Lehman Brothers	-6	-7	9	8
Bailouts: October 3, 2008	-6	-6	9	9
AIG	-6	-6	6	8
Other banks	-6	-3	9	9
Tax issues	2	0	7	8
Healthcare/ so security	2	1	7	7
Congressional debates	1	0	5	7
Public anxiety	-6	-3	9	8

Table 3.4: Qualitative Ratios: Obama Campaign:
One Dramatic Moment Attribute - Raw and Smooth Scores

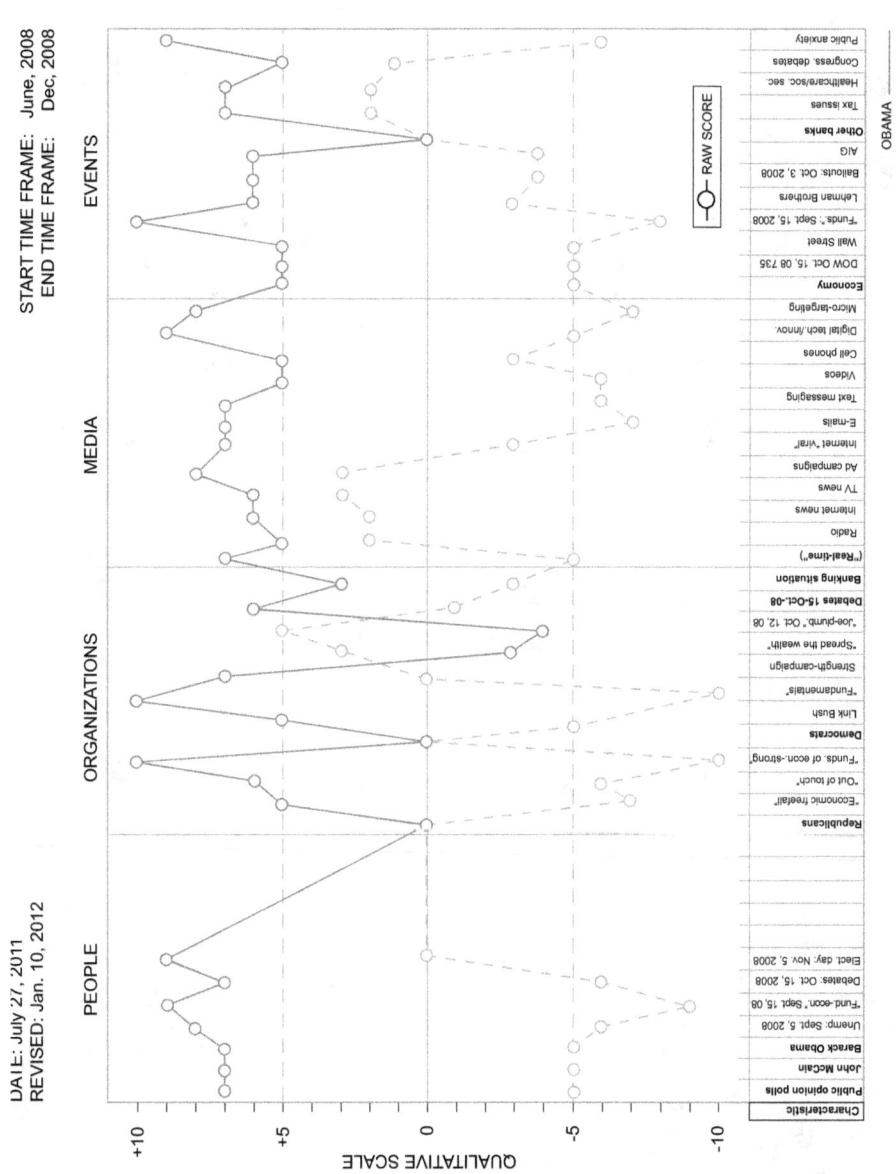

Chart 3.4 (r): Obama Campaign: One Dramatic Moment Attribute -
Raw Scores

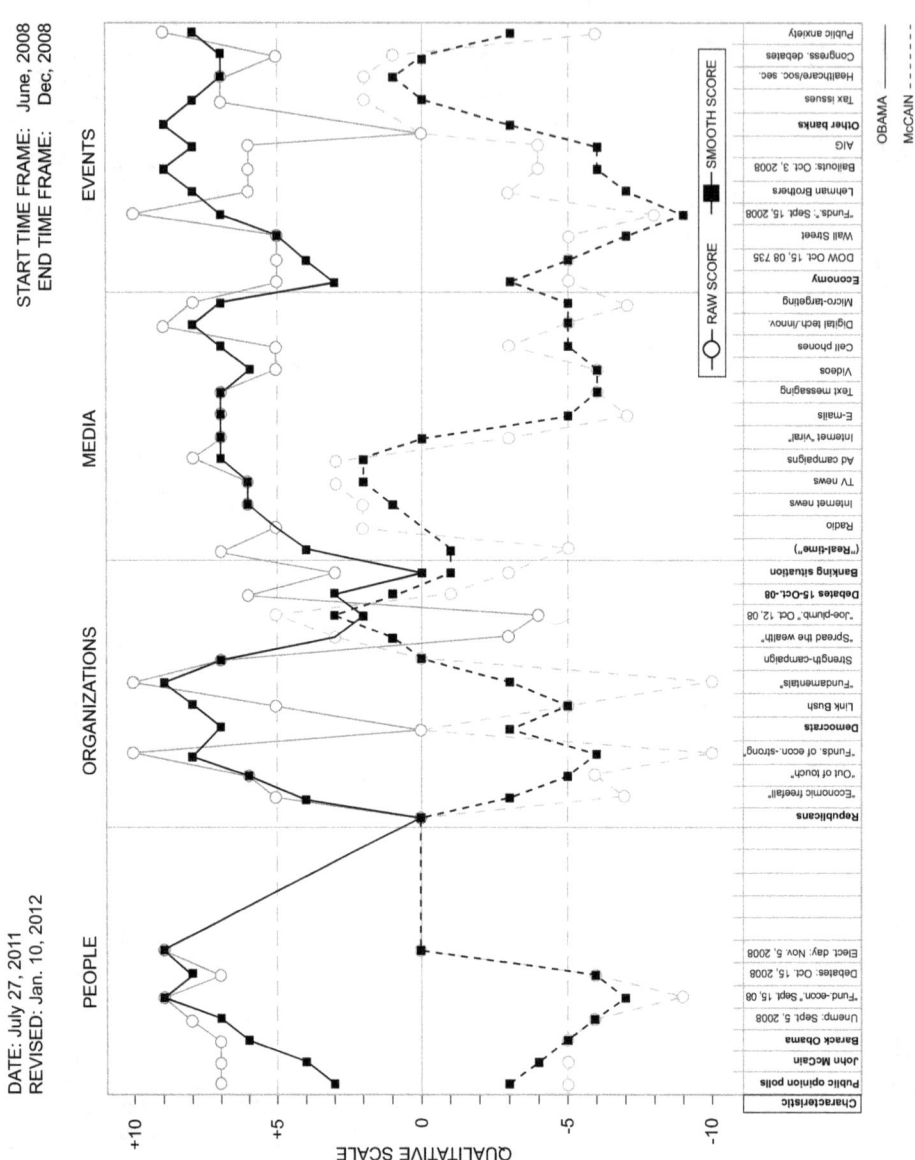

Chart 3.4 (rs): Obama Campaign: One Dramatic Moment Attribute -
Raw and Smooth Scores

Section 3.5: Obama Presidential Campaign Compilations

On November 4, 2008, Barack Obama became the first African-American to be elected president of the United States. He gained almost 53% of the popular vote and 365 electoral votes. The popular vote percentage was the best showing for any presidential candidate since George H. W. Bush in 1988 and not since Bill Clinton, whose 379 votes in 1996. He won Colorado, Nevada, Virginia, Indiana, Florida, Ohio, and North Carolina, all states that were won by President George W. Bush in 2004. A bylaw received more total votes than any presidential candidate in history, totaling well over 69 million votes.

In response to the first research question:

Did the media influences have a direct impact on the presidential hopeful Barack Obama victory?

From these overwhelming results one can see that the *media* coverage of the economy issue provided the necessary *momentum* and *stickiness* to propel Obama to his *tipping point* on or about the middle of September 2008. In addition, Joe Biden also made history by being the first Roman Catholic to be elected vice president, and after serving in the Senate for 36 years prior, was the longest serving senator to become vice president.

With regard to our other important question:

Was there a direct correlation between the media frequency and content of the messages during the campaign?

The Obama campaign raised enormous amounts of money that was spent throughout the media sectors on delivering messages to the voters and contributed to the outcome. Future campaigns will copy the Obama approach by offering similar forms of messages that can be proliferated on the Internet almost immediately creating a communication vehicle capitalizing on the interactive character and potential of future media delivery systems. (As evidenced in 2012, during the Obama vs. Romney US Presidential Campaigns)

Candidates who are willing to subscribe to the new *media* platforms and to spend (and find investors to spend for them to raise more cash) to increase and expand their *stickiness momentum* largely will have the advantage against candidates who are unwilling to please prospective patrons and utilize these advantages in the same way.

For example, big oil companies which had an influence on previous campaigns' strategies, may actually hurt future candidates rather than help sway voter opinion due to the associated negative impacts and links to the environment related to fossil fuels and *climate change*.

In contrast, another possibility is the better-financed candidate might micro-target (saturate) his audience and reach voters without the benefit of rebuttal from the other candidate, which as the social platforms currently illustrate, is risky, as in a truly democratic societies the opportunity to defend one's position and one's agenda is paramount and would be canceled out. Rather than electing a better person with the better policies and better agenda, the American voters might be influenced by only one candidate who has the ability to use (*flood*) the media with enough micro-messages that distort the candidates' positions and the real potential for democratic solutions. With the media developing more in *real-time*, such strategies may prove detrimental. Nevertheless, these point out some interesting complexities as to how media forms in this context are influential.

As we turn to the second case study, the *International Financial Crisis 2007– 2010*, similar patterns in the *contagiousness* and *stickiness attributes* are prevalent, and ultimately, the one *dramatic moment tipping point* is impacted by many of the same types of *media* communication delivery systems that delivered the messages and contributed to the Obama Presidential Campaign *epidemic*.

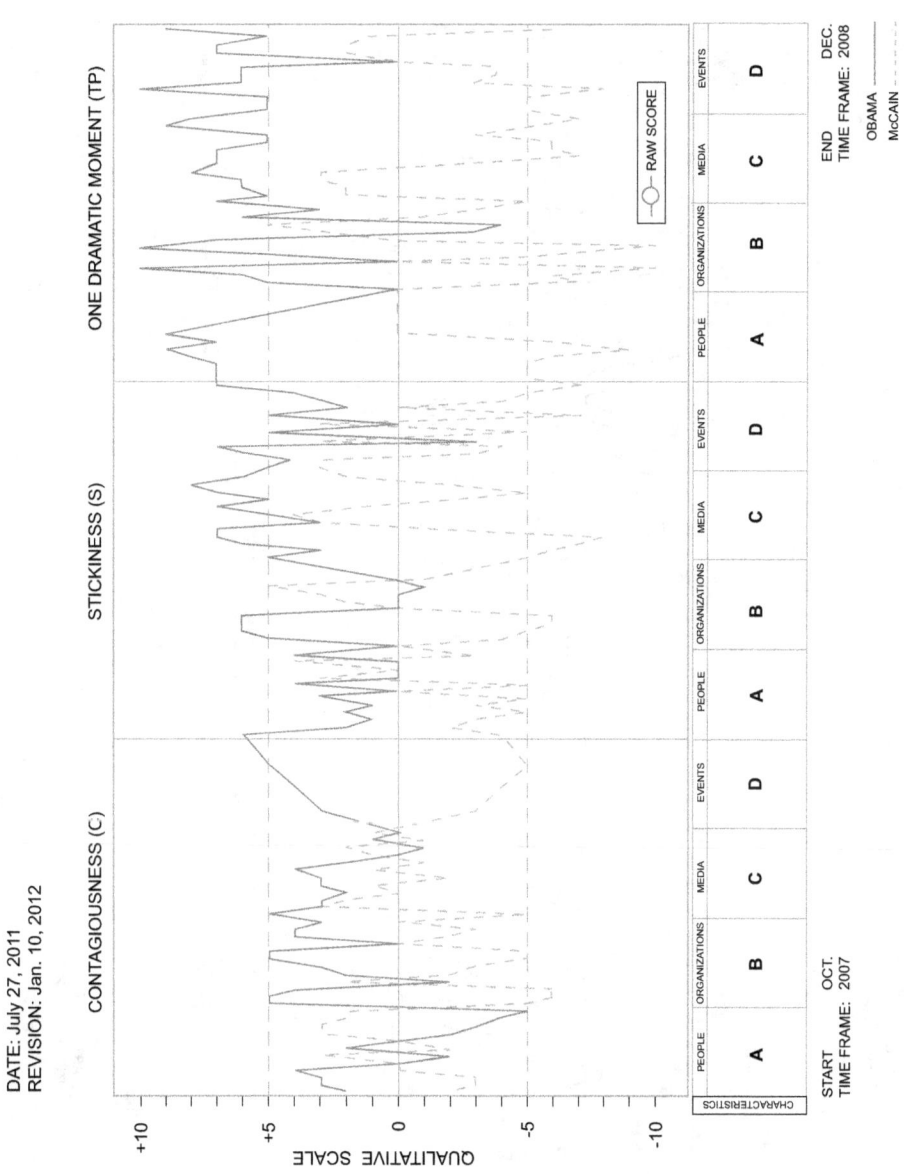

Chart 3.5 (r): Obama Campaign: Compilation: Raw Scores

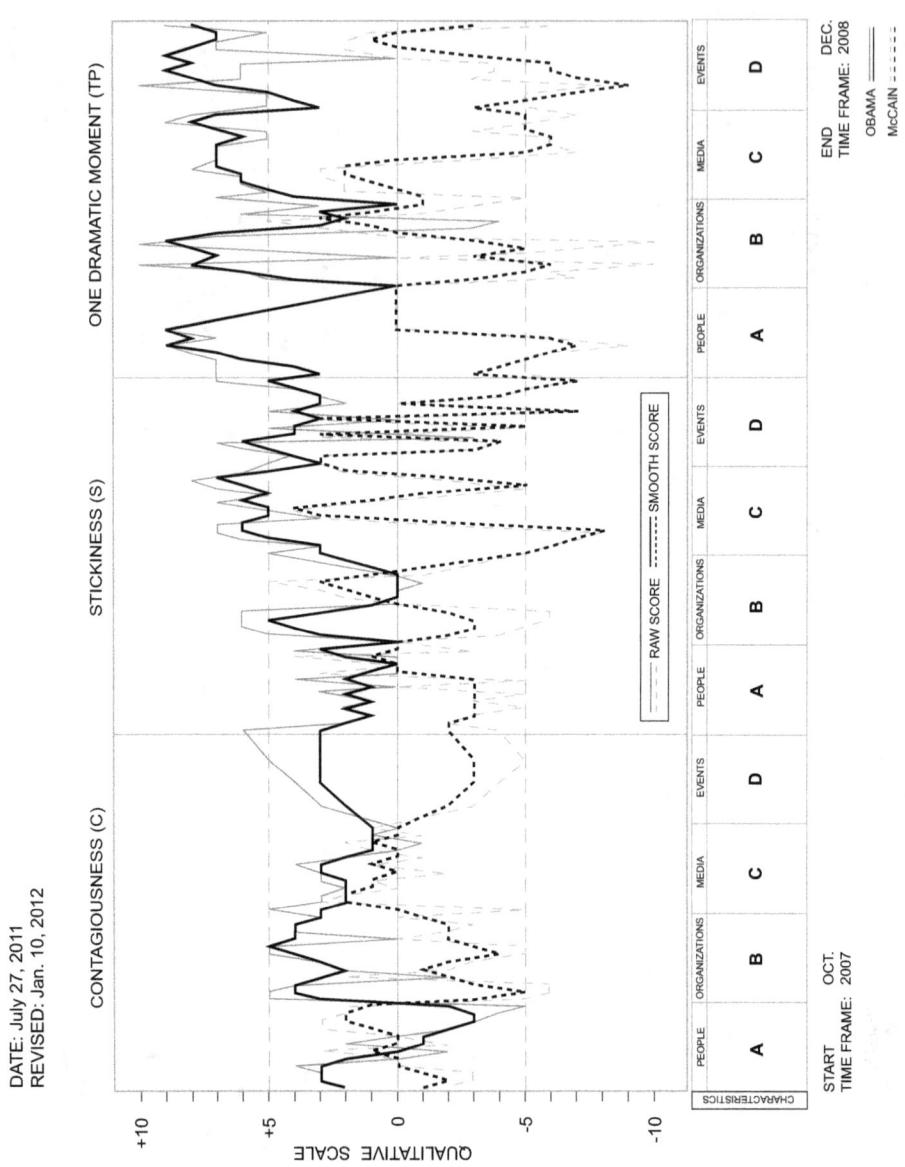

Chart 3.5 (rs): Obama Campaign: Compilation: Raw and Smooth Scores

CHAPTER 4: INTERNATIONAL FINANCIAL CRISIS 2007–2010

Section 4.1: Introduction

The context for the research that was formulated for this chapter about the second case study, the *International Financial Crisis 2007–2010* (it is now 2013), demonstrates that uncertainty continues to create nervousness over the future of the global and American economies.

Now tsunamis, radioactive plumes, Middle East revolutions, the ongoing European Debt Crisis and a still weakened United States economy could undermine the recent gains. Some global issues, like the spike in oil and food prices, are quantifiable. But the long term, clearer picture depends on indicators yet to come, and recent history only highlight some of the *attributes* and *characteristics* of the complexity of *global financial events* and *media influence.*

"The problem is not Japan alone—it's that Japan reinforces all the negative repercussions and our own weak recovery," said Stephen S. Roach, non-executive chairman of Morgan Stanley Asia and a professor at Yale. *"It's difficult to know the tipping point for the global economy, but there are difficult headwinds now."* The sequence of recent *global events* adds to a sense of global foreboding.

Recently in Libya, American warplanes were flying and the oil producing wells stood silent. Troops from Saudi marched into Bahrain, moving across the Persian Gulf from Iran. And in Europe, with the Greek crisis temporally adverted, finance ministers warn that hundreds of banks in other countries still carry billions of dollars in bad loans.

As surveys of global economists by *The International Economy Magazine* presented a majority view it as likely that some combination of Greece, Ireland and Portugal will likely default on debt and force investors to take heavy losses. Oil prices continue to rise as of this version. Japan, already the largest importer of liquefied natural gas, searches for energy to replace a damaged nuclear grid, as analysts expect these prices to rise too.

Finally, the United States, with an economic colossus burdened by the *International Financial Crisis 2007–2010* which resulted in the worst long-term unemployment situation in nearly a century. If Japanese companies and investors retrench, selling some Treasuries and investing fewer yen overseas, the pain here could grow.[103]

In the years leading up to the start of the *2007–2010 International Financial Crisis*, large amounts of foreign money flowed into the U.S. from rapidly-growing economies in Asia and oil-producing countries. These funds made it easy for the *Federal Reserve* to hold interest rates in the United States extremely low from 2002–2006 which contributed to easier credit conditions, which culminated in the United States housing bubble. Various types of loans (e.g., mortgage, credit card and auto) were extremely easy to obtain whereby consumers assumed an unprecedented debt load.[104] Additionally, as part of the housing and credit booms, financial agreements called mortgage-backed securities (MBS) and collateralized debt obligations (CDO) proliferated, these derived their value from mortgage payments and housing prices. These financial instruments enabled institutions and investors throughout the world to participate in the U.S. housing market. As housing prices fell, major global financial institutions which had borrowed and invested heavily in subprime (MBS and CDO markets) reported significant losses. Falling prices also lead to homes being worth less than the mortgage, and providing owner incentives to option for foreclosure. The foreclosure *epidemic* that began in late 2006, drained substantial wealth from consumers and eroded the strength of banking institutions. Similarly, defaults and losses on other loan types also increased as the crisis spread from the housing market to other parts of the economy. Total losses estimated in the trillions of U.S. dollars globally.[105]

With the housing and credit bubbles, we researched how the tipping point *attributes* and *media* involvement influenced the financial sector to both

103 Michael POWELL, "Crises in Japan Ripple Across the Global Economy," *New York Times*, March 20, 2011. Retrieved from http://www.nytimes.com/2011/03/21/business/global/21econ.html?pagewanted=all
104 Paul KRUGMAN, "Revenge of the Glut," New York Times, March 1, 2009. Retrieved from http://www.nytimes.com/2009/03/02/opinion/02krugman.html?_r=1
105 "Executive Summary," *International Monetary Fund*, January 2009. Retrieved from http://www.imf.org/external/pubs/ft/weo/2009/01/pdf/exesum.pdf

expand and ultimately, to become increasingly fragile. Policy makers failed to recognize the increasingly important role played by financial institutions, particularly, investment banks and hedge funds, (also known as the shadow banking system). Some experts believed these institutions became as important as commercial (depository) banks in providing credit to the U.S. economy, however, not subject to the same regulations.[106] These institutions as well as certain regulated banks had assumed enormous debt burdens providing the loans (described above) and did not have a financial reserves sufficient to absorb large loan defaults or MBS losses. Resulting in these losses impacting the ability of financial institutions to lend, diminishing economic activity. These concerns drove central banks to provide funds to stimulate lending and try to stabilize the commercial paper markets. Governments responded and provided bail outs to key financial institutions and assumed significant additional financial commitments.

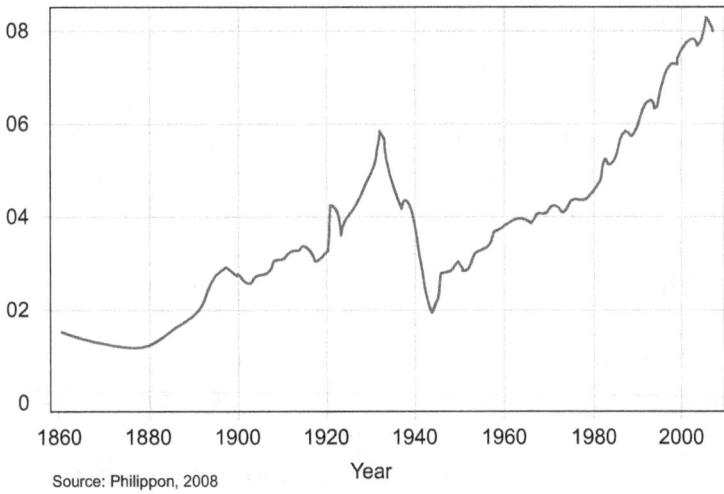

Source: Philippon, 2008

Figure 4.1: Share in GDP of U.S. financial sector since 1860[107]

From the period between September 1, 2007 and November 30, 2010, the world markets slipped into one of the worst financial meltdowns since the 1930's, coupled with complete loss of equity value and confidence in the banking system worldwide. The inevitable impact (tipping point) worldwide was the one week *casualty list* (October 18, 2008), including *Lehman Brothers*, *AIG Insurance, Morgan Stanley* acquiring *Citibank*, and *Washington Mutual*, the federal bailout of *Bear Sterns*, and the immediate infusion of the 800

106 Timothy GEITHNER, "Reducing Systemic Risk in a Dynamic Financial System," Speech, *Federal Reserve Bank of New York*, June 9, 2008. Retrieved from http://www.newyorkfed.org/newsevents/speeches/2008/tfg080609.html
107 Thomas PHILIPPON, "The Future of the Financial Industry," *Finance Department of the New York University Stern School of Business at New York University, Stern on Finance blog*. Retrieved from http://sternfinance.blogspot.com/2008/10/future-of-financial-industry-thomas.html

billion dollar rescue package by United States government and related global stimulus packages throughout the world. The financial crisis was carried, *blow-by-blow*, with updates on almost all *media* as a "*living story*."

Recalling the *Broken Windows Theory*, the brainchild of James Q. Wilson and George Kelling, which argued that, in effect, *impacts* and *events* are the inevitable result of *disorder*.[108] A simple application of this *phenomenon* suggests the key research question for this study:

If a "*window*" is broken and left unrepaired, i.e. the international financial system, whereby, if it appears that the system is unregulated, unmanaged or, that no one cares about it and no one is in charge, soon more windows will be broken, and the condition will spread (*contagiousness*), sending a signal that anything goes?

In countries or regions, illustrating this example, financial oversights and lack of monitoring regulation are all the equivalent of *broken windows*, and promote invitations to more serious events (as is evident in recent events in Ireland and Greece that have affected the entire European Union).[109] *This is the epidemic of finance*. It equates to that the effects are *contagious* - just as a *trend* is *contagious* - that it can start with just one broken window, or country financial failure, and spread instantaneously, particularly if the message is delivered via media networks to an entire region's population and the world economies. The impetus to engage in a certain type of behavior is not coming from a certain kind of person or culture, but from a feature of the environment, i.e. financial sector – banking. *Contagiousness* postulates that if a particular behavior in a region or community (or world) goes unaddressed, it signals that nobody cares about it resulting in additional behavior of the same type.[110]

The *Tipping Point Theory attributes* are presented here in the context of this *global event* and *media impact*, and the relationship to the panic and sell-offs, the unprecedented drop in the United States stock market of over 700 points in one day, and similar losses in markets worldwide with a net value loss reaching 10 trillion in USD.

Whereas sometime in October 2008, the world financial markets finally got the message. The world was riveted into a vicious credit crunch downward *contagiousness*, and it had reached the brink of frightening proportions.

---

108 James WILSON and George KELLING, "Broken Windows. The Police and Neighbourhood Safety," *Atlantic Magazine*, 3, 1982.
109 Ibid.
110 Ibid.

Stock markets worldwide plunged, currencies experienced violent ups and downs, and lending between banks stopped. Governments, in turn, poured trillions into bailout loans, equity infusions, and massive interventions. *The United States Federal Reserve Bank*, in a never experienced before move and unprecedented expansion of the powers, released 1.1 trillion USD of new lending in a period of about six weeks. This tsunami of money went to banks, to big insurers, to commercial paper users, and to money market funds. A 700 billion dollar bailout bill flew through American Congress on the promise that it would get at the root cause of the American crisis and purchase the toxic assets from the banks' ledgers. European governments, led by Great Britain, attempted to outdo the American bailout and focus their infusion directly into the banks.

For the first time in history, finance ministers worldwide realized how lethal the new financial instruments created and brokered in the United States had saturated the global investment portfolios, and to what extremes the banks, especially in Europe, had gone in imitating the American corporate giants. A fashionable word in Europe was it could "*decouple*" its economy from the Americans, but this strategy quickly vanished as the European continent crept towards negative growth during the same time. The oil producing states- Russia, Venezuela, Iran, and the Arab states--that had linked their economies to the American consumer were faced with the same problems. Emerging economies—like Korea, Taiwan and Brazil—were staggering and caught up in the epidemic of the melt down. One country in particular, Iceland, which had taken a much riskier path, went bankrupt. The global crisis was indeed *Made in America*, despite imitators and those that followed its path. At its core it was a consumer binge on imported goods by the class of super rich who invented nothing and built nothing except for intricate chains of paper that people mistook for wealth.

Looking at all these *events* from a tipping point perspective, the late spring of 2007 seems like a totally different era. The American financial markets were robust, consumer spending was continuing to grow, the market for investment grade credit was booming, and insurance premiums demanded for those who wanted to invest in riskier forms of debt were at an all-time low. Case in point, the S&P 500 jumped more than 9% just from March through May of 2007. One of the first contagiousness events came about in mid-June of 2007 occurred it was learned that two *Bear Sterns* mortgage hedge funds could not meet their margin calls. At that point, *Moody's* downgraded some of their subprime mortgage-based bonds. The fund sold some of its bonds to raise money, and the rest were just not salable at any price. This was the first time that subprime-related debt tumbled. While the experience was a wake-up call, financial analysts tried to remind the world that subprime mortgages were a small percentage of the overall financial market and that the problem had "*been contained*." This was the one of the first contagiousness *living stories* carried in the media, and it became the lead story and propagated (*stickiness*) throughout all of the media forms of distribution networks.

However, the *contagiousness* started to spread all around the world related to subprime related funds, i.e. when a 9 million dollar London hedge fund closed its door, there was a run on the big London mortgage lender, German and Swiss banks announced large write-offs, and in August of 2007, the *Federal Reserve* and the *European Central Bank* began to flood their economies with money. It was now a global problem and the *media* was all over it in huge proportions. What followed were additional revelations from big banks, especially *Citigroup*, announced they held hundreds of billions of long-term loans and mysterious off-balance-sheet entities called "*SIV's*" that they had financed in the short-term commercial paper market. The shock of these kinds of disclosures brought interbank lending to a screeching halt. In addition, other banks revealed they were holding billions in bridge loan commitments to finance highly leveraged private equity company buyouts.

The *stickiness* continued as the *Federal Reserve* came to the rescue and aggressively cut the base short-term lending rate in September and another in October 2007. At that juncture, the Stock market and the credit markets "*shuttered*" back to life. The problem was that the loss that was disclosed in October was shocking - 20 billion in asset write-downs, at *Merrill Lynch* and *Citigroup*, but the problem was, they had grossly underestimated their losses to a point that in November, analysts were told they just did not know the value of these instruments that were at the center of their problem. The October 2007 fiasco was the first of many *events* which would follow in subsequent quarters as losses at major banks kept spreading along with the uncertainty about the real value of the bank assets.

Following a series of *contagious and stickiness events*, CEOs were fired, *Federal Reserve* interventions were more extreme, and in December 2007, the Fed tried to re-liquefy banks, exchanging treasuries for some of the riskier credit instruments. Through the spring of 2008, the *epidemic* expanded and continued as the *Federal Reserve* increased the instruments that they would accept as collateral for the loans to companies, but the attempts to turn back this rising tipping point virus, as it was now infecting the financial systems, became less and less successful. As the impact was felt worldwide, nervous markets teetered on the edge of the impending *tipping point*.

The first big bank to topple was *Bear Stearns* in March of 2008. Like most of the large investment banks, its trading ledgers were highly leveraged and depended on short-term financing. And because doubts spread about its overall portfolio, lenders finally refused to roll over its credit lines. Ultimately, bankruptcy was avoided only by a forced merger with *J.P. Morgan.* The *events* continued. *Countrywide Financial*, the largest American mortgage lender had to be rescued by the *Bank of America* in May. And in August, *Fannie Mae* and *Freddie Mac*, the two giant mortgage lenders with some 5 trillion dollars in home loans, were taken over by the US government. *Lehman Brothers*, which was bigger than *Bear Stearns*, was in worse shape, and with no merger prospects, *Lehman* filed for bankruptcy on September 15, 2008.

This was the beginning of the *tipping point*, or the *one dramatic moment* in the international financial crisis. That very same weekend, the insurance giant *AIG*, petitioned the Fed for a large temporary loan and was rejected like *Lehman*, yet it was not even a bank. *AIG* was a guarantor of 300 billion of American mortgage-backed CDOs held by European banks. If *AIG* failed, the European banks would be forced to write off some hundred and 50 billion in assets. *The Federal Reserve* had to re-posture with a $85 billion loan, which grew to $123 billion after some very difficult negotiations. Next *Merrill* saw the handwriting on the wall and merged with *Bank of America*. And that same week *Morgan Stanley* and Goldman Sachs petitioned the Fed to convert to full *Federal Reserve* status, basically surrendering their freedom from regulation for the assurance of federal aid should they need it. As the crisis *spread* throughout the world markets, the US announced its $700 billion bailout plan. And at that point, all European governments entered the markets in force. By November of 2008, one would have been hard pressed to find a major bank on the continent that did not receive a large infusion of government cash.

In 2013 at the time of this writing, while there was a soft rebound after 2009 into 2010, the stock market continues to be fragile (with wild valuation swings), and the credit and lending markets remain semi-catatonic, particularly in Europe. This tipping point case study and the analysis of what happened, was not just a banking *epidemic*, as we will discover. The issues went much deeper than that.

The approach for this case study was to analyze the *media* influence during the *International Financial Crisis 2007–2010* in two key areas:

The first, with regard to the *contagiousness and stickiness attributes,* was there a *"living story"* in the *media* that contributed to the systemic international financial crisis?

Second, which relationships between the *attributes* and *characteristics* and various *media* delivery systems led to the historic international financial crisis tipping point?

Extracting from the research (*Media Tipping Points*), we analyzed each *attribute* with the goal of identifying the *characteristics* and understanding the relationships between them as the *2007–2010 International Financial Crisis* unfolded.

With the research questions above and the analysis methodology, what follows are the tipping point *qualitative ratios* and *graphic summary* of how the *attributes* and *characteristics* were influential relative to *people, organizations, media and events*. It *identifies* and *quantifies* the relative impact and relationships among them. This analysis of the *attribute* relationships evolves and become more important in context of the impact on the outcome. The same series of rules governed the research and provided a means of control over what *characteristics/issues* were included as part of the research. The influence of the *media* and the messages delivered by and with the technology that led to the tipping point for the *International Financial Crisis of 2007–2010* was presented. The influence of the *media* and the messages delivered by and with the technology that provided for the *tipping point* for the *International Financial Crisis of 2007–2010.*

Clearly, one can interpret these *characteristics/issues* and *attributes* from many perspectives; however, the intent of this *tipping point analysis* is to illustrate *why* and *when* the *one dramatic moment event* occurred during the International Financial Crisis and its origins and then compare these in *Chapter Six* to the other two case studies. This method for analyzing the *attributes* is simply to look at the *sequence of events* in a chronological manner and observe the way the international financial crisis developed and how the various *media* sectors played their role, ultimately contributing to the worst global financial meltdown since the 1930s.

Section 4.2: Contagiousness Attribute (C)

Date:	05-Sep-11	Start Time Frame:	January 2007
Revised:		End Time Frame:	January 2008

International Financial Crisis: Contagiousness Attribute

Qualitative ratio scale	Raw score	Smooth score
Characteristic		
People		
Baby boomers	5	3
Manufacturing loss	2	2
Untrained workers	2	1
Politics	0	0
Culture wars	-2	-1
Nixon	-3	-2
Carter	-2	-1
Reagan	2	1
Clinton	4	3
Bush	6	4
World leaders	3	3
Bank directors	2	2
Organizations	0	0
Banks	3	3
Auction rate/security system	4	4
Bank runs	5	5
Wall Street	5	5
LBOs	6	5
Deregulation	4	4
PIKS/derivatives	3	4
Technological instruments	6	5
Trading tools	6	5
Internet revolution	7	6
Gov. organizations	3	5
Private sector	4	4
Media	0	0
"Living story"	5	3
Conservatives	1	4
Liberals	7	5
Bank risk	5	5
Subprime mortgages	6	5
Housing	5	5
Bush	6	5
Obama	1	4
Big business failure	7	5
Bank sector	7	6
Stimulus package/tarp	1	5
US auto industry	7	3
Events	0	0
Bank failures	6	3
Stock indexes	7	5
Equity value	5	5
Credit defaults	6	5
Liquidity	6	6
Trade deficits	6	6
Banking shakeout/subprime lending	8	7
US financials/GDP	5	6
European financials	5	5
US reserves	3	4
"Warnings ignored"	8	6
CDOs	5	3

Table 4.2: Qualitative Ratios: International Financial Crisis:
Contagiousness Attribute - Raw and Smooth Scores

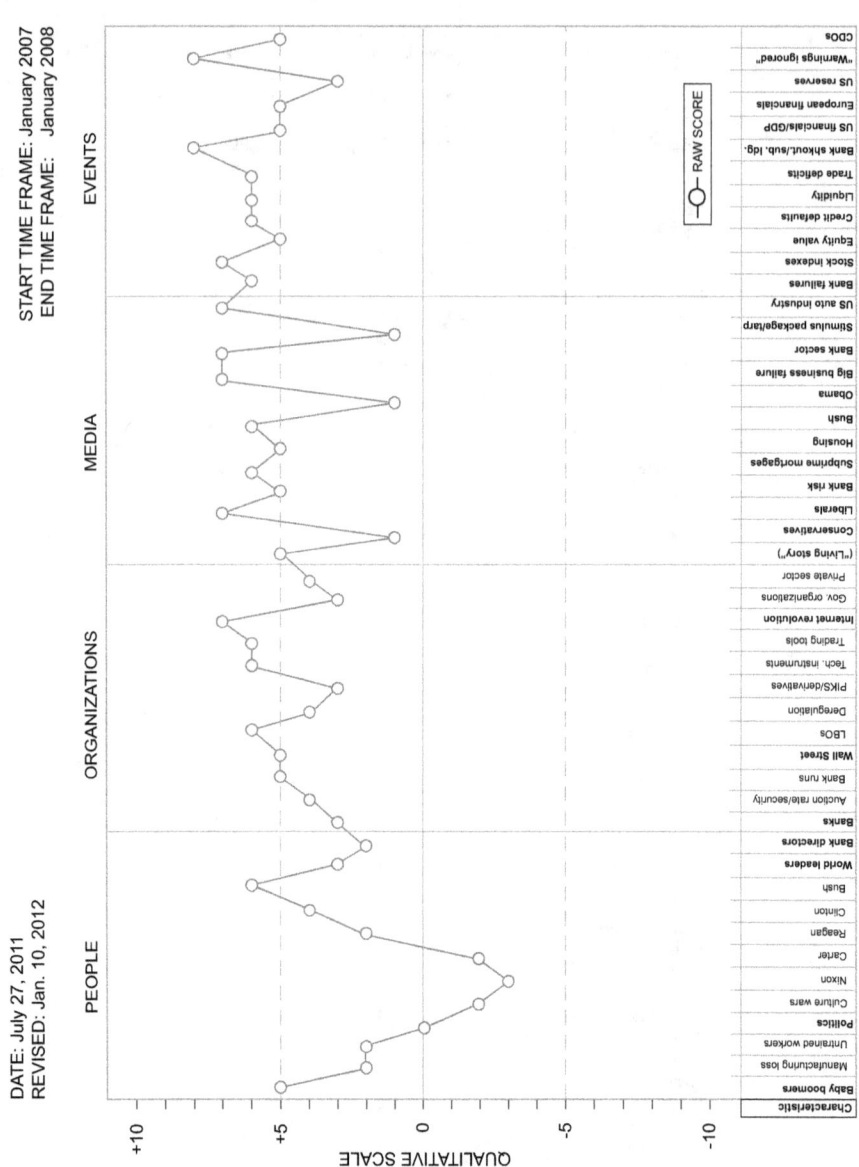

Chart 4.2 (r): International Financial Crisis:
Contagiousness Attribute - Raw Scores

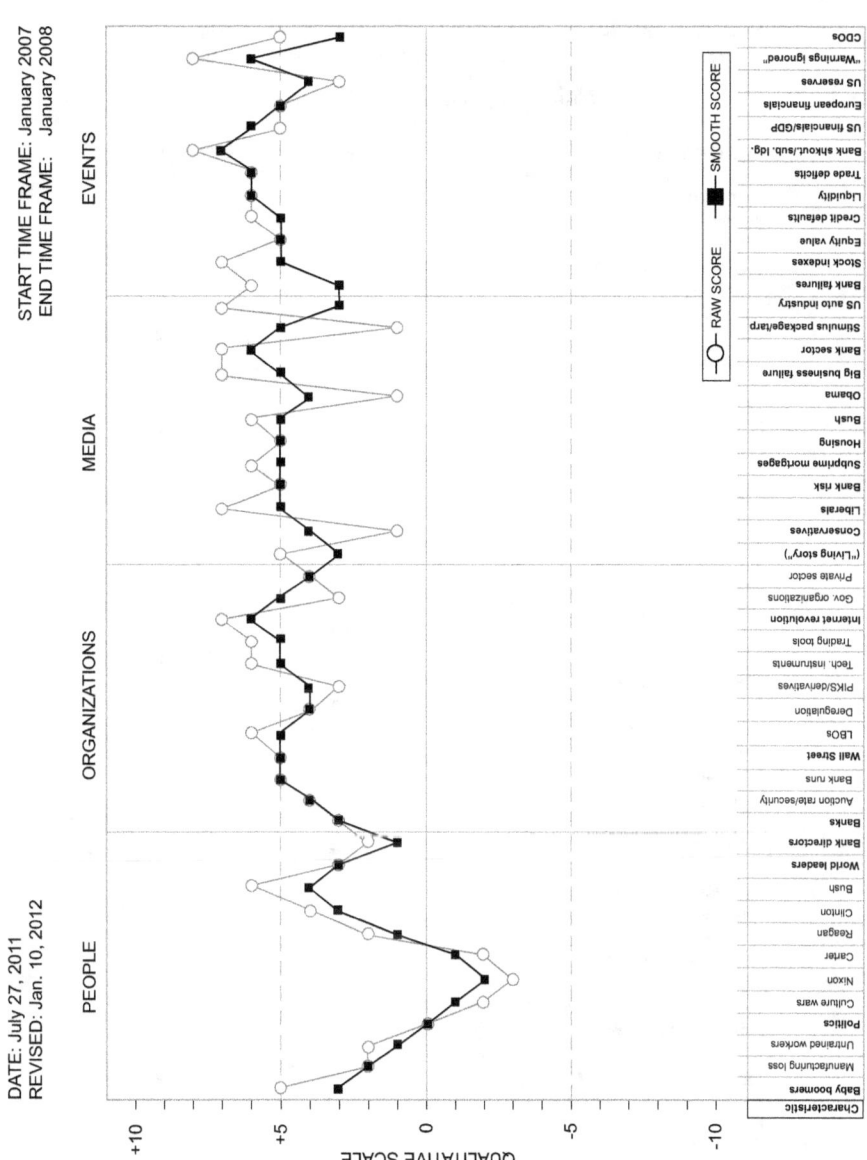

Chart 4.2 (rs): International Financial Crisis:
Contagiousness Attribute - Raw and Smooth Scores

Section 4.3: Stickiness Attribute (S)

Date:	05-Sep-11	Start Time Frame:	January 2008
Revised:		End Time Frame:	January 2009

International Financial Crisis: Stickiness Attribute

Qualitative ratio scale	Raw score	Smooth score
Characteristic		
People	0	0
Alan Greenspan	3	2
Federal Reserve/FMOC	1	4
American property owners	5	4
American borrowers	5	3
Others	0	2
Henry Paulson	4	3
Benard madoff	3	3
Timothy Geithner	2	2
Robert Gibbs	1	2
Ben Bernanke	2	1
Organizations	0	0
Banks		
Citibank	7	3
Merrill Lynch	7	5
Other World Bank failures	8	7
Subprime lending	9	8
Application process	6	6
Credit scoring	6	6
Financial instruments	8	7
Mortgage-backed securities	8	4
Media	0	0
"Livingstory"	9	5
Banking/subprime/defaults	8	7
Commercial paper/loans	7	7
Mental recession?	9	8
Public reaction	9	6
Competition with other events	2	4
Presidential campaign	3	3
Iraq war	2	2
Hurricanes, wildfires	1	1
Time delay/"easier story to tell"	-5	0
Gas prices	5	3
Public attention/New York/Washington DC	4	2
Events	0	0
Recession	6	3
Housing defaults	7	5
Gas prices	6	6
Crisis in banking	9	7
Bailouts	9	8
Stimulus package	3	7
US auto industry	8	7
Subprime lending/Fannie Mae and Freddie Mac	7	6
Government actions	5	6
Media themselves	9	8
Financial innovation/risk-taking/"gaussian copu	9	7
Elections	5	4

Table 4.3: Qualitative Ratios: International Financial Crisis:
Stickiness Attribute - Raw and Smooth Scores

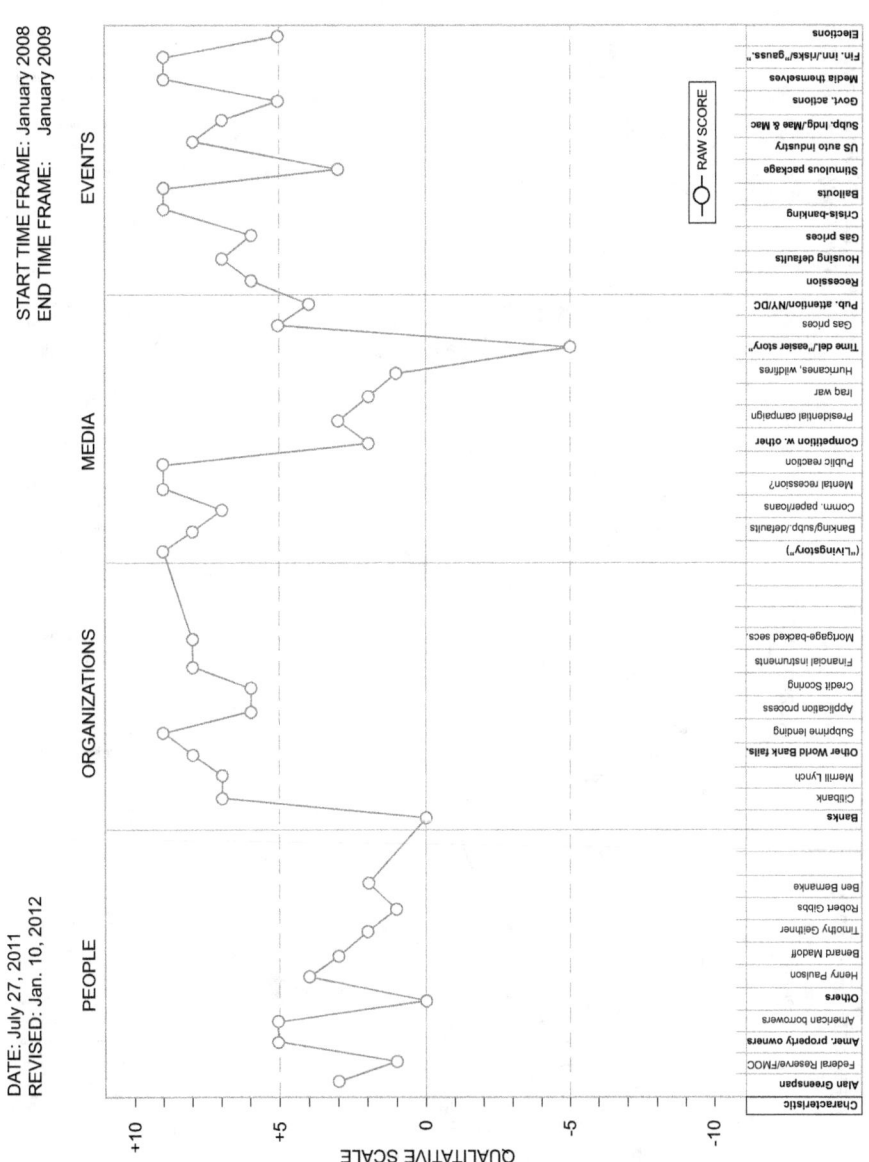

Chart 4.3 (r): International Financial Crisis: Stickiness Attribute -
Raw Scores

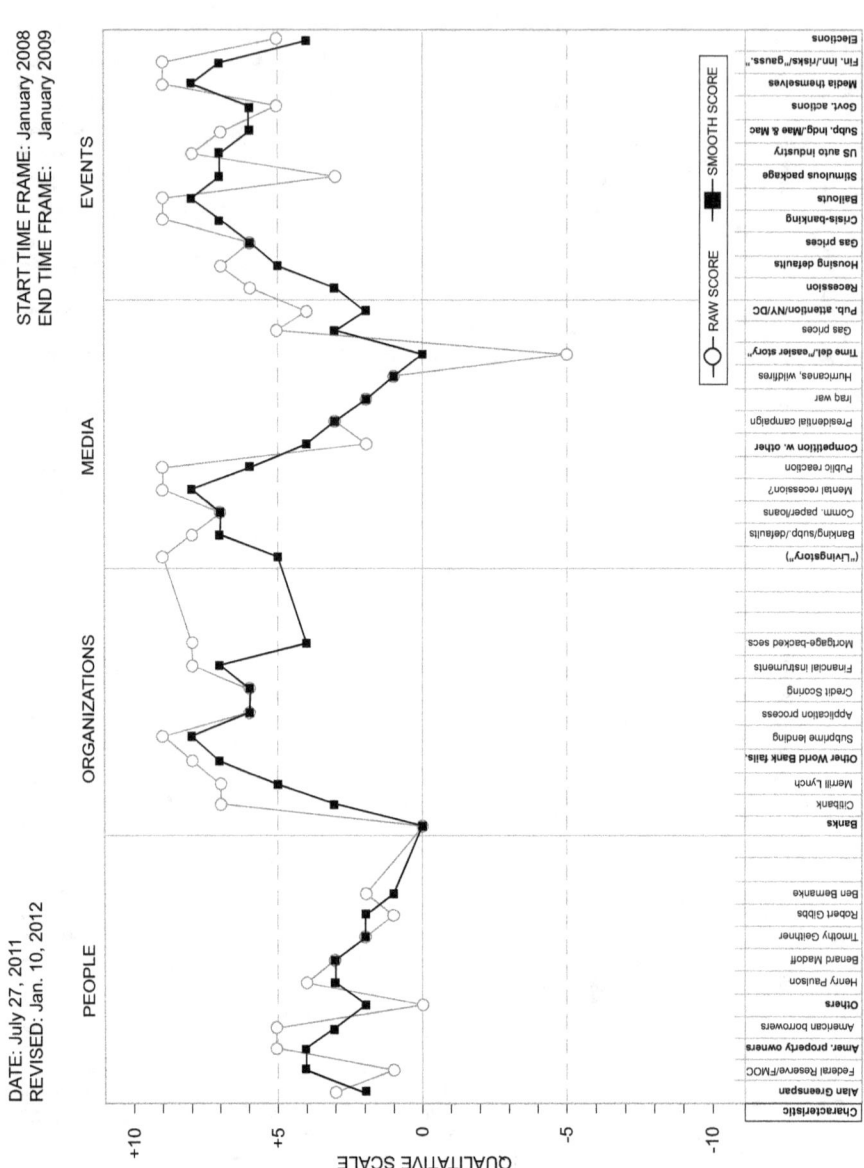

Chart 4.3 (rs): International Financial Crisis: Stickiness Attribute -
Raw and Smooth Scores

Section 4.4: One Dramatic Moment: The Tipping Point (TP)

Date:	05-Sep-11	Start Time Frame:	January 2009
Revised:		End Time Frame:	January 2010

International Financial Crisis: One Dramatic Moment Attribute

Qualitative ratio scale	Raw score	Smooth score
Characteristic		
People	0	0
Alan Greenspan	3	2
Timothy Geithner	1	3
Bank directors	6	4
Regulators	-1	4
Government officials	4	4
Bush	7	5
Obama	1	3
Others	0	1
Donald Trump	1	1
Warren Buffett	2	1
Organizations	0	0
Shadow banking	9	5
Derivatives	9	7
Off sheet balances	9	8
Credit default swaps CDS	9	9
Over-the-counter OTC	9	9
Deregulation	8	8
Securities and exchange	7	8
Stock markets	9	9
DOW	10	7
Congressional debates	4	6
Bailouts	-2	3
Stimulus package	-3	2
Media	0	0
"Living story"	10	5
"Bad news versus good news"	10	7
"Warnings ignored"	9	8
Market rebound	5	5
Competing stories -healthcare, terrorism	4	4
Bailouts/stimulus package	3	5
Media influence	9	7
Newspapers	4	6
Radio	5	7
TV	9	8
Websites	7	8
Viral Internet/blogs	9	8
Cell phones/video	7	4
Events	0	0
"Denial thinking"	7	3
"Social virus"	8	6
"Rippling effect"	9	7
"Information cascade"	9	8
Subprime disaster	10	9
Financial institutions	9	9
Housing "myth"	8	9
Stock markets	9	8
Debates/stimulus package	4	6
Government reports	5	7
Banking crisis	10	8
Global market reaction/impacts	9	4

Table 4.4: Qualitative Ratios: International Financial Crisis:
One Dramatic Moment Attribute - Raw and Smooth Scores

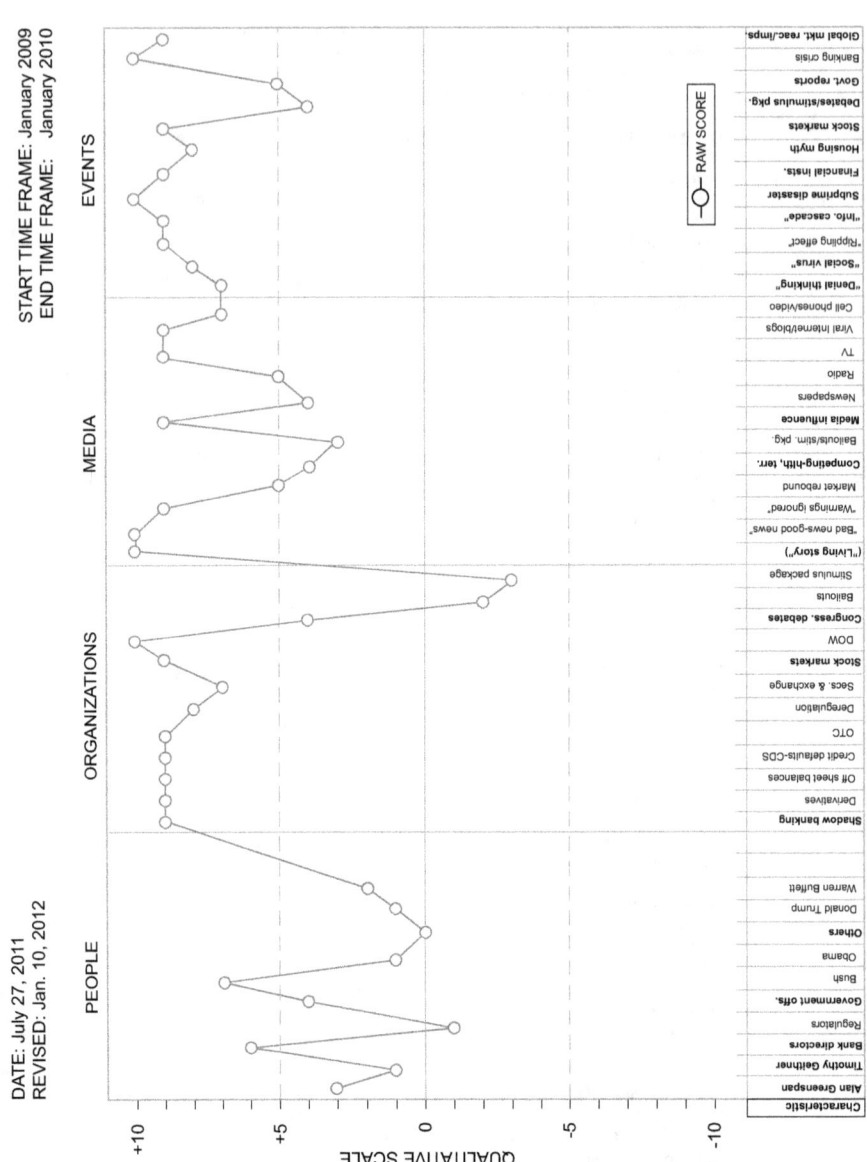

Chart 4.4 (r): International Financial Crisis:
One Dramatic Moment Attribute - Raw Scores

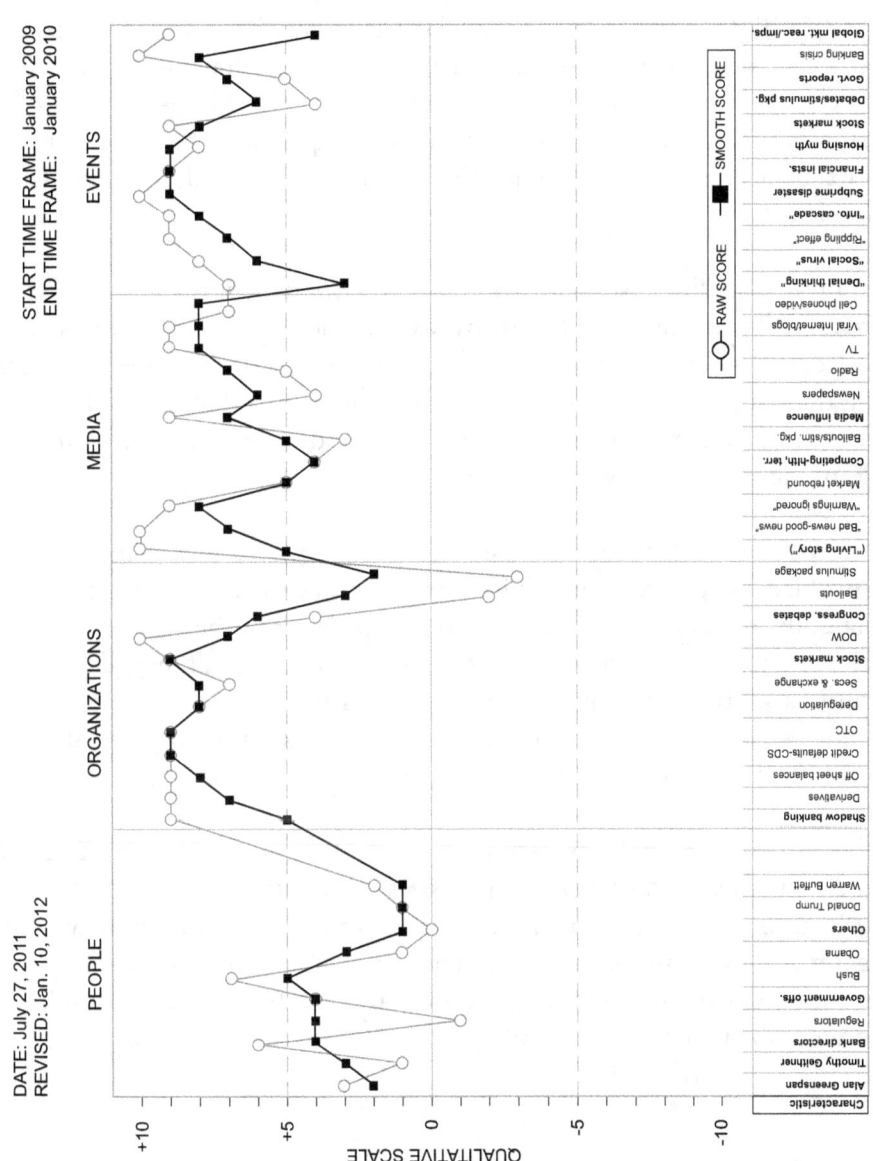

Chart 4.4 (rs): International Financial Crisis:
One Dramatic Moment Attribute - Raw and Smooth Scores

Section 4.5: International Financial Crisis 2007–2010 Compilations

The 2007–2010 International Financial Crisis has been called by leading economists the worst financial since the Great Depression of 1930s with declines in consumer wealth estimated in the trillions of US dollars contributing to the failure of key businesses, requiring substantial financial support by governments, and leading to a significant decline in overall economic activity.

In response to our first research question, this *contagiousness* and the *stickiness* "*living story*" was being delivered by worldwide *media* that influenced the credit freeze that brought the global financial systems to the brink of collapse during the last quarter of 2008. The central banks introduced $2.5 trillion of government funds, the largest injection of liquidity into the credit markets and banking systems in world history.

The convulsion continued with the media delivery systems during the crisis as it profiled the global housing bubble, which peaked in the United States during 2006 and chronicled and *spread* the *epidemic* as the value of complex securities tied to housing prices to plummeted and damaged financial institutions globally. It impacted the global stock market, which suffered large losses during 2008. Economies worldwide slowed in late 2008 and early 2009 as credit was tight and international trade declined.

As we turn to our final case study, *Climate Change*, similar patterns are prevalent in the *contagiousness* and *stickiness attributes*, and ultimately, could lead to the world's *one dramatic moment*, or, *tipping point*, which is characterized by many of the same types of *media* communication delivery systems that delivered the messages and that contributed and influenced the outcome in the first two case studies.

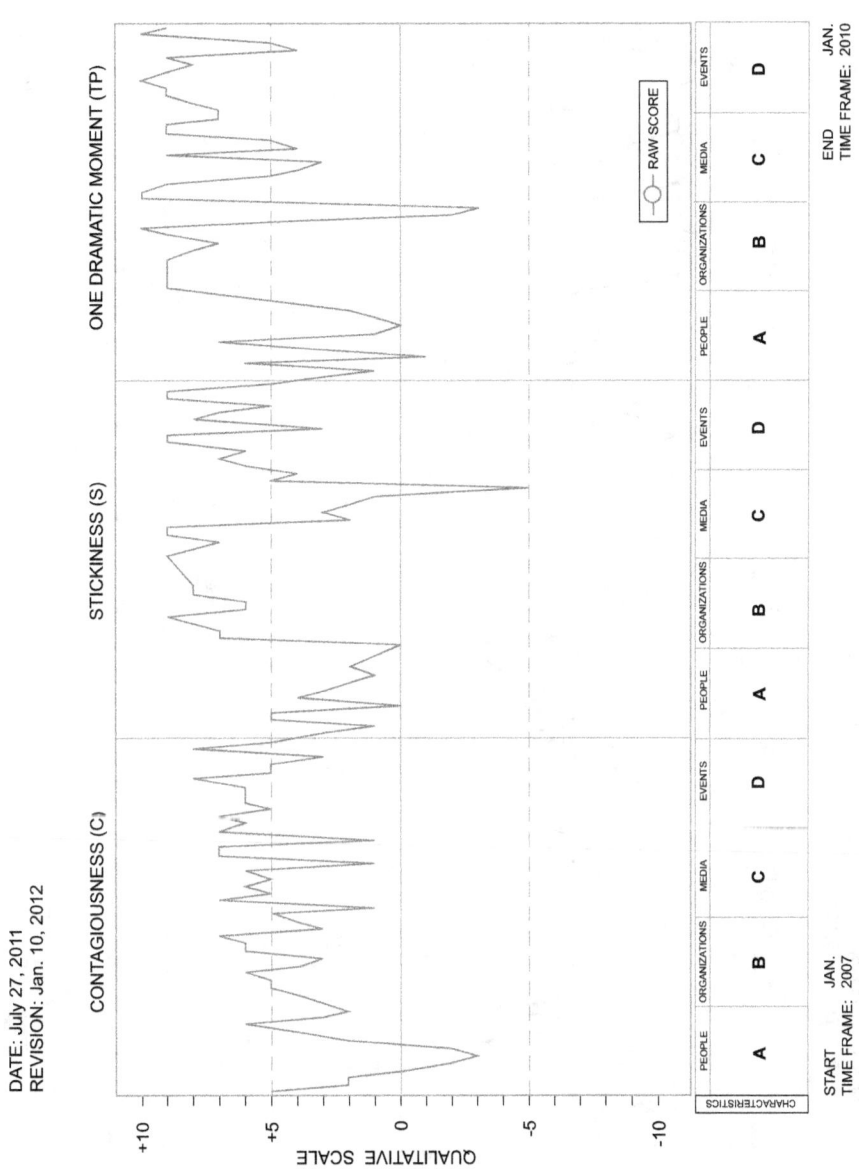

Chart 4.5 (r): International Financial Crisis: Compilation: Raw Scores

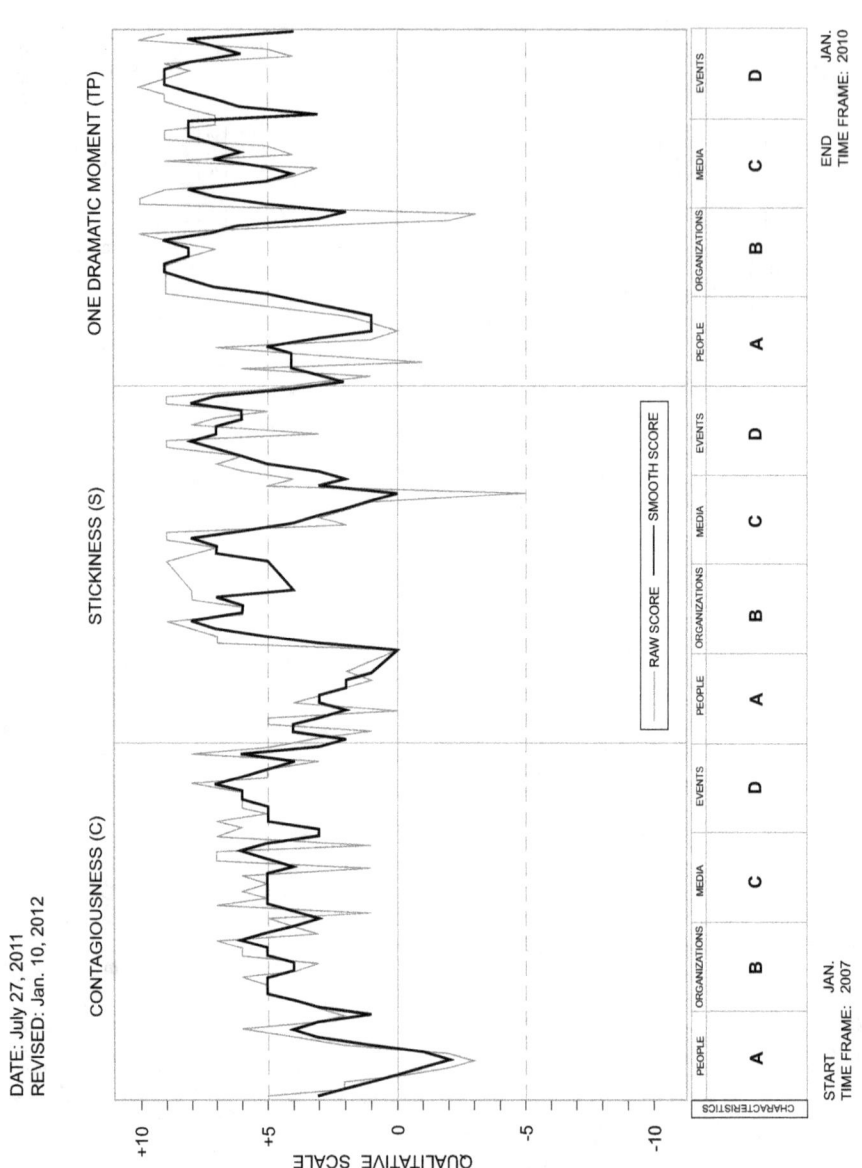

Chart 4.5 (rs): International Financial Crisis:
Compilation: Raw and Smooth Scores

CHAPTER 5: CLIMATE CHANGE

Section 5.1: Introduction

Much has been said recently about the increasing need to fuel our cities and economies and the toll it is taking and effect on our planet's energy resources. This last case study explored some of the political, economic, cultural, and moral issues relevant to *Climate Change* as *global event*.

From the onset of this case study I would like to begin by framing the discussion:

Science is about probability, not certainty. And the persisting uncertainties in climate science leave room for argument. What is a realistic estimate of how much temperatures will rise? How severe will the effects be? Are there tipping points beyond which the changes are uncontrollable?[111]

In approaching this study, this expectation was to find a blistering controversy about *Climate Change* with no clear position. However, the research pointed out an overwhelming consensus, and even though mainstream media was pretty much on the same page, doubt had worked into in all the reports about *Climate Change*, as for every scientist warning of global warming, there was another scientist saying that the claims were inaccurate. Countless scholarly reports are cited herein where essentially all of the world's leading scientists and academics appear all to be speaking with the same voice. They all say the same thing...that the world's climate is changing dangerously, and humans are to blame.

---

111 Andrew C. REVKIN and John M. BRODER, "Facing Skeptics, Climate Experts Sure of Peril," *New York Times*, December 7, 2009, p. A1, A8.

The research questions became:

"Why had climate change become a media debate?"

Media and technology had influenced the other two case studies and their outcomes to a large degree.

Was perhaps the murkiness or confusion on the climate change issue brought about by an organized media campaign as well?

Or was it financed by the biased agendas, such as those of the coal and oil industries, to make us think that climate change science is somehow still controversial, that climate change is still unproven?

This case study presents two *attributes* only, the first, concerning the *contagiousness attribute*--identifying research, trends, and facts that exist-to-date. The second, explores the stickiness attribute—and presenting how the system of *sharing of the truth* has clouded the issue and pointing out some of the media impact on this *living story*.

My personal objective is a *"redirect"* towards a *global consciousness.* We also explore whether or not it is possible that this new *global consciousness,* fostered and promoted by *media* platforms, could be focused to address our future global reality with new technologies (and those which are still evolving) in order to avoid a catastrophic *global event.* Finally, the study highlight the scenarios for a *"best case"* and *"worst case"* of the potential tipping point, or *one dramatic moment* for *Climate Change.*

The intention was more than just to present a research collection of articles, conclusions, and opinions. Rather, in my earlier book, *Media Tipping Points,* there is a complete narrative which puts in context the *tipping point attributes* and *characteristics* similar to that of the *Obama Presidential Campaign* and the *International Financial Crisis* studies.

This third case study illustrates that *Climate Change* as a *global event* is directly related to *media impact* and is certainly a conversation that needs to be had, as we have arrived at a *critical juncture* in human history. By mastering our new (and innovative) media technologies and utilizing the media influence for our *global consciousness,* we can explore whether or not we are capable of outperforming the inevitable and changing the course of history, in effect, to remake the global environmental landscape and rewrite the future.

Two important points:

The debate about *Climate Change* is mostly conducted by people who live in cities where everything is artificial. Where they don't actually experience the changes that are taking place in the real world, and,

The scientists, who have been telling us the truth when they say the world is at risk, and if we listen to those who are manipulating the media, we are in trouble.

This third case study, *Climate Change,* was fundamentally focused overall on analyzing each of the *tipping point attributes* and *characteristics* similar to the first two case studies in order to identify those that can be understood for the future, and be predicted. It became evident that complex *global media systems* which are engaged and manipulated, ultimately, could influence this case study outcome. The approach for this case study was to analyze the *media environments* and *technology* available and their relationships to *Climate Change* and then to answer these research questions:

Does media contagiousness and stickiness attributes influence the world's behavior as it relates to potential climate change?

What are the relationships between these attributes and characteristics and various media environments that may lead to a tipping point in climate change?

Can media influence and predict and/or avoid a potential world tipping point from occurring with Climate Change, rather, can complex global media systems be engaged and executed with a "global consciousnesss" to influence the outcome?

Extracting from the research (*Media Tipping Points*), we analyze here the *attributes* with the goal of identifying the *characteristics* and understanding the relationships between them as they occur. With the research questions above and the analysis methodology, what follows are the tipping point *qualitative ratios* and *graphic summary* of how the *attributes* and *characteristics* were/ and are influential relative to *people, organizations, media and events.* It *identifies* and *quantifies* the relative impact and relationships among them.

This study of the *attribute* relationships evolves and become more important in context of the impact on the outcome. The same series of rules governed the research and provided a means of control over what *characteristics/*

issues were included as part of this analysis. The influence of the *media* and the messages delivered by and with the technology were included that may lead to the tipping point for the *Climate Change* is presented. The influence of the *media* and the messages delivered by and with the technology that provide an opportunity for the tipping point application.

Clearly, as with the other studies, one can interpret these *characteristics* and attributes from many perspectives; however, the intent of our *tipping point analysis* is to define where and when the *one dramatic event* may occur and then compare these in *Chapter Six* to the other two case studies. The goal is to identify the actions and reactions of the various media sectors along the *Climate Change* journey and identify the potential relationships of these *attributes*. Our future generations demand this and similar approaches.

Section 5.2: Contagiousness Attribute (C)

Date:	05-Sep-11	Start Time Frame:	January 2010
Revised:		End Time Frame:	January 2030

Climate Change: Contagiousness Attribute

Qualitative ratio scale	Raw score	Smooth score
Characteristic		
People		
7 billion people	5	5
Human activity	7	4
Scientists	1	3
Industry Agendas	6	3
Government leaders	1	1
Corporate lobbyists	-2	0
University "think tanks"	1	1
Al Gore	2	0
Pres. Bush	-3	-3
Organizations		
IPCC	2	1
Wikileaks	0	3
Energy industries	8	6
Oil	7	5
Coal	8	6
Nuclear	5	5
Politics	1	2
US	3	3
Religion	4	3
Special-interest groups	-2	0
Environmental groups	2	2
Scientific organizations	3	1
Media		
"Harder story to tell"	-5	-1
"age of mis-information"	-4	-2
Oral culture	2	0
Internet distortion	-3	-1
TV and cable TV	1	0
Radio	1	1
Print and newspapers	0	0
Viral blogs textos e-mails	2	0
Standard of accuracy	-4	-2
Government campaigns	-1	-1
Political pressure on scientists	-3	-3
Media cover-up?	-5	-3
Events		
Greenhouse effect	5	3
1990 to 2100 : 1.4° to 5.8°	7	5
Climate models	5	6
Adaptation	-1	3
Health effects/disease	1	2
Sea levels	2	1
Ability to change	0	0
Risk aversion / denial	-2	-1
Migration	1	0
Resources	2	1
Conflict	5	3
Valuation of ecosystem	4	4

Table 5.2: Qualitative Ratios: Climate Change:
Contagiousness Attribute - Raw and Smooth Scores

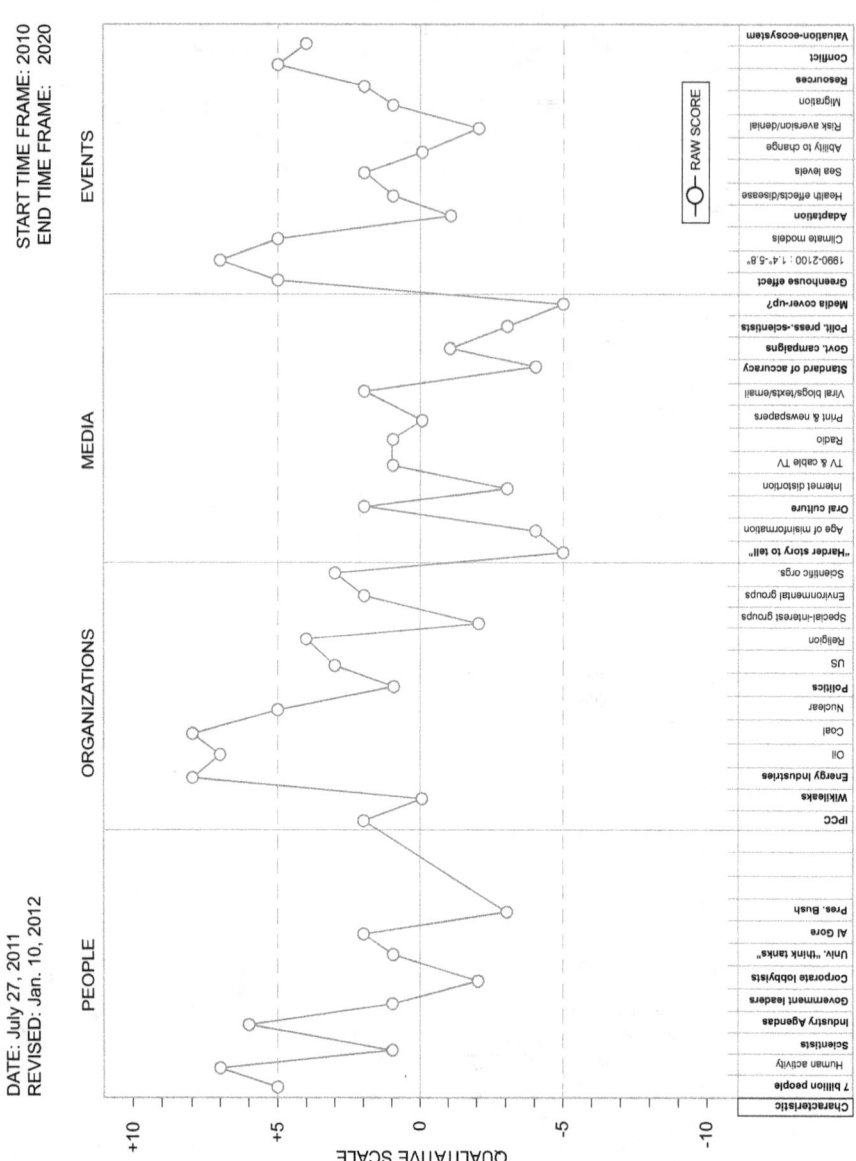

Chart 5.2 (r): Climate Change: Contagiousness Attribute - Raw Scores

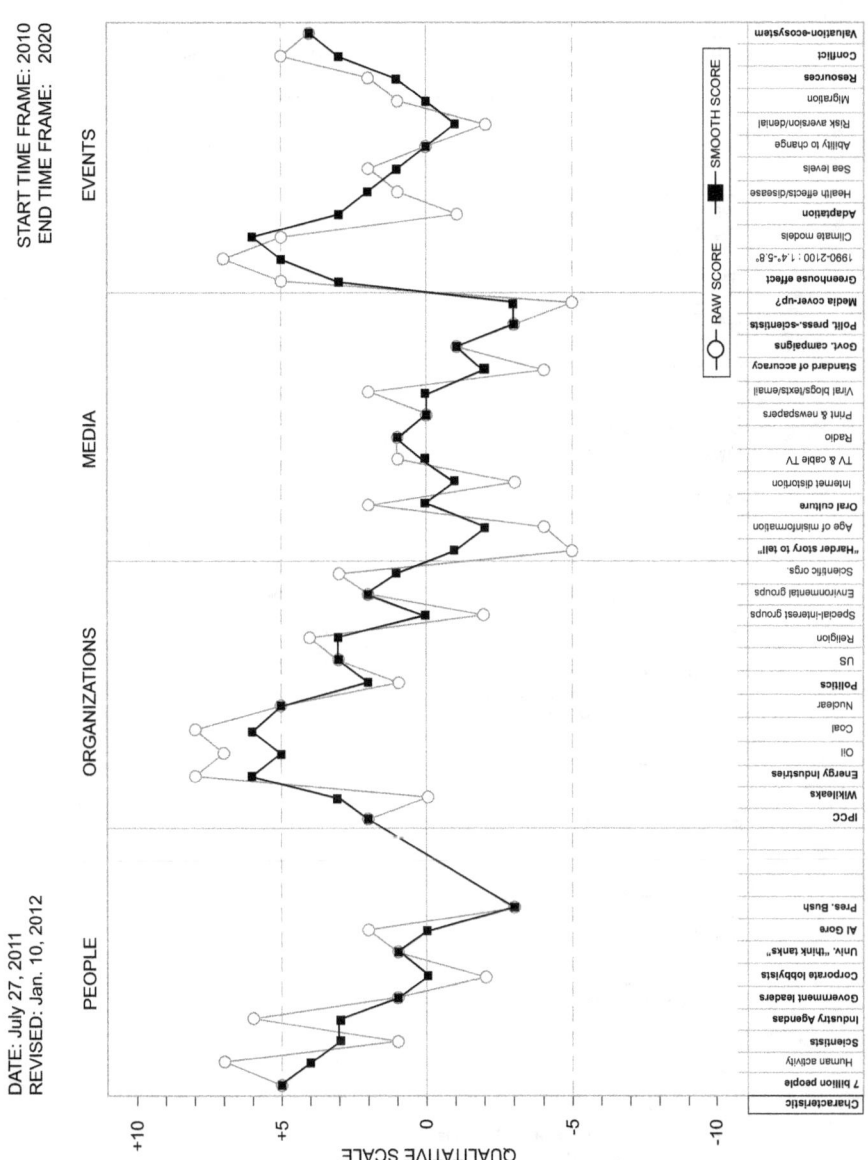

START TIME FRAME: 2010
END TIME FRAME: 2020

DATE: July 27, 2011
REVISED: Jan. 10, 2012

Chart 5.2 (rs): Climate Change: Contagiousness Attribute -
Raw and Smooth Scores

Section 5.3: Stickiness Attribute (S)

Date:	05-Sep-11	Start Time Frame:	January 2030
Revised:		End Time Frame:	January 2100

Climate Change: Stickiness Attribute

Qualitative ratio scale		Raw score	Smooth score
Characteristic			
People			
Citizens of the planet		6	3
CO2 levels		8	4
Scientists		0	4
Keeling	Keeling curve	2	2
Miles R. Allen		1	1
Joseph Cannadel		1	1
George Bush	1992	-2	0
Al Gore	2006	2	1
Populations		6	4
Developed countries		7	5
Undeveloped countries		6	4
Organizations			
US Department of energy		1	1
Bob Watson		1	1
IPCC		3	0
Bush administration	Congress 1998	-3	-1
Obama administration		0	0
Kyoto		1	1
Copenhagen		1	1
United Nations		2	2
World meteorological organizations		2	2
National Academy of Sciences		2	2
Al Gore		3	2
Other scientists		2	1
Media			
"Real-time"		5	4
"Living story"		6	5
"Harder story to tell"		9	7
Distorted media space		8	7
Doubts		7	6
"Climategate"		2	3
Internet		3	3
Blogs texts e-mails		3	3
Television and cable TV		1	2
Newspapers		1	1
Radio		1	1
Geo technical solutions		0	0
Events			
Temperature change: 5-6°F or 18° F versus 3.6° F		8	6
CO2 up 25% by the year 2020		9	7
Adaptation		3	4
Droughts		5	5
Floods		5	5
Food scarcity		5	5
Capital Investments		1	3
Economies		2	5
Quality of life		9	7
Conflict		8	4

Table 5.3: Qualitative Ratios: Climate Change:
Stickiness Attribute - Raw and Smooth Scores

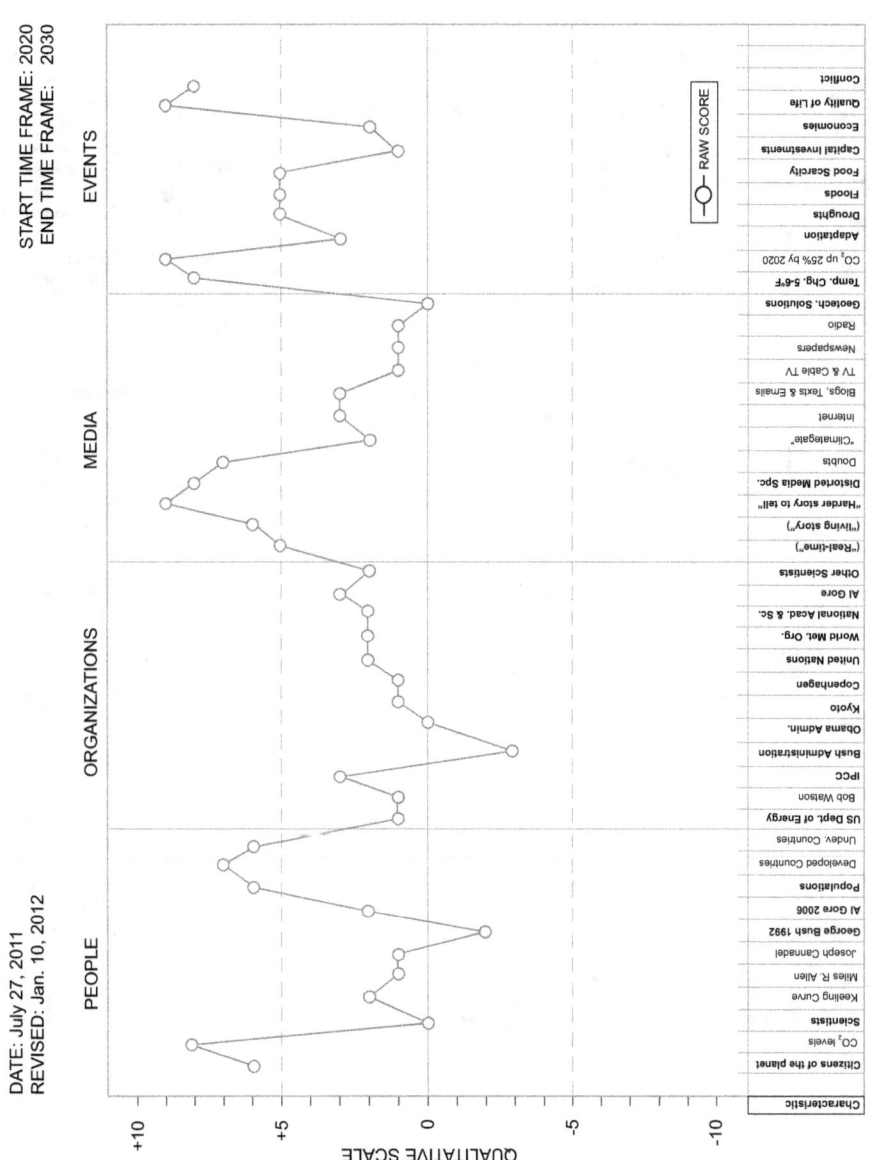

Chart 5.3 (r): Climate Change: Stickiness Attribute -
Raw Scores

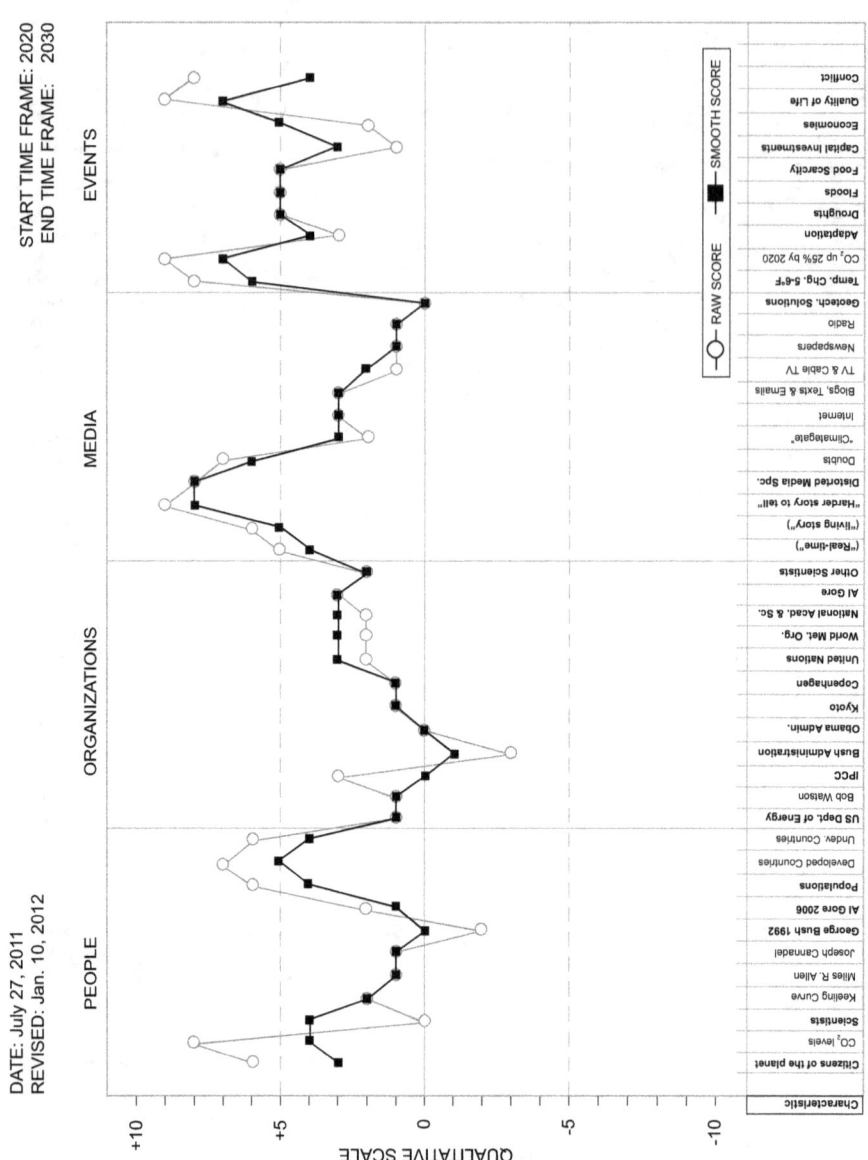

Chart 5.3 (rs): Climate Change: Stickiness Attribute -
Raw and Smooth Scores

Section 5.4: One Dramatic Moment: The Tipping Point (TP)

Date:	05-Sep-11	Start Time Frame:	January 2100
Revised:		End Time Frame:	January 2140

Climate Change: One Dramatic Moment Attribute

Qualitative ratio scale	Raw score	Smooth score
Characteristic		
People		
World Impacts	6	5
Population	7	6
Resources	5	5
Air	6	6
Water	6	6
Food	7	6
Shelter	5	5
Other	6	6
War/ conflict	8	5
Organizations		
Generations of leaders		
Forefathers	-1	0
Present	-4	-1
Future	0	0
International borders	2	1
USA	2	2
European Union	3	2
Arab states	2	1
South America	0	0
Asia	-5	-2
African Union	-6	-3
New World environmental groups	0	-1
Media		0
"Global consciousness"	7	3
Social media platforms	8	6
Other		5
Government intervention	4	5
Media events	8	7
Food/ energy prices	7	7
Population	8	7
Hurricanes/tornadoes/floods/ droughts	7	7
Displaced regions	7	7
Topple governments	9	8
"Denial thinking"	6	7
Media management	2	3
Events		0
Climate change revolution	9	5
"Global consciousness"	10	7
Other		
"Crisis driven change"	10	8
Scenarios		
Best case		
Year 2100		
Events:New York, Venice, Singapore, etc.	10	9
Worst-case		
Year 2140		
Social unrest, war, chaos, 36.4 billion people	10	5

Table 5.4: Qualitative Ratios: Climate Change:
One Dramatic Moment Attribute - Raw and Smooth Scores

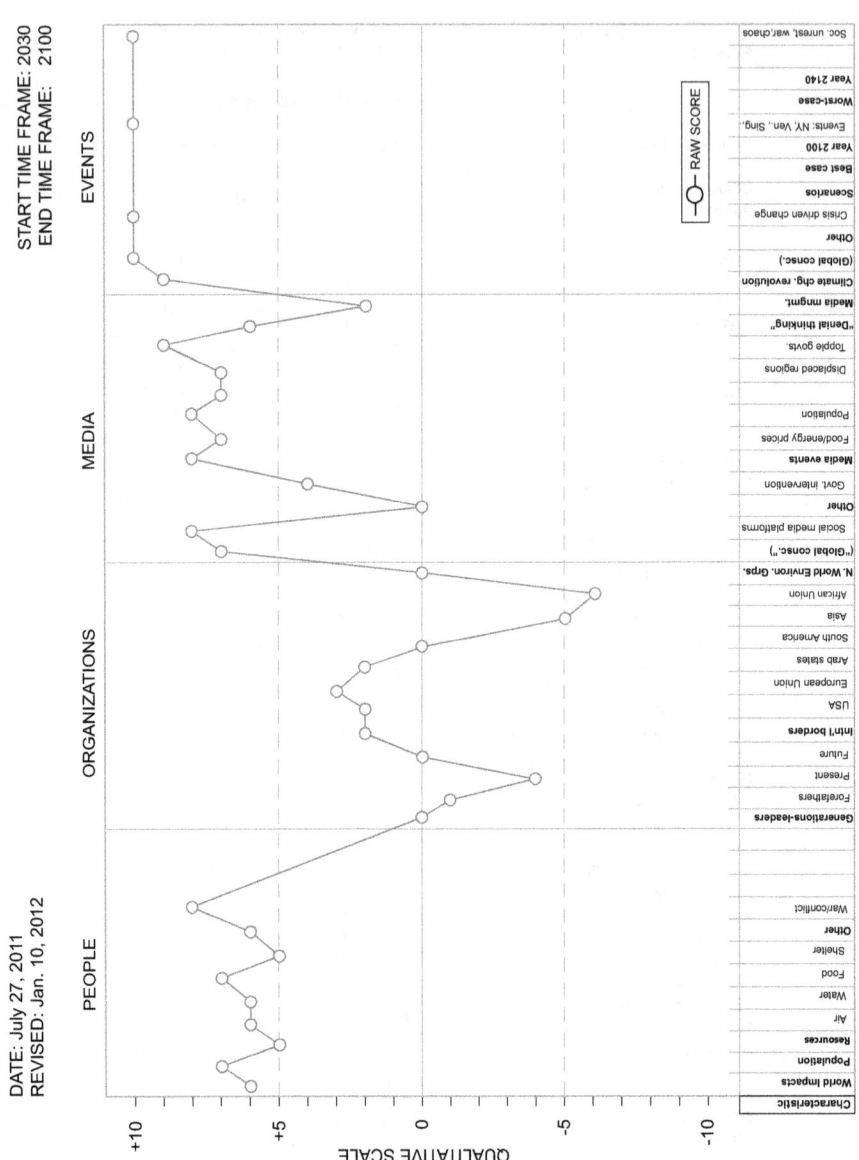

Chart 5.4 (r): Climate Change: One Dramatic Moment Attribute -
Raw Scores

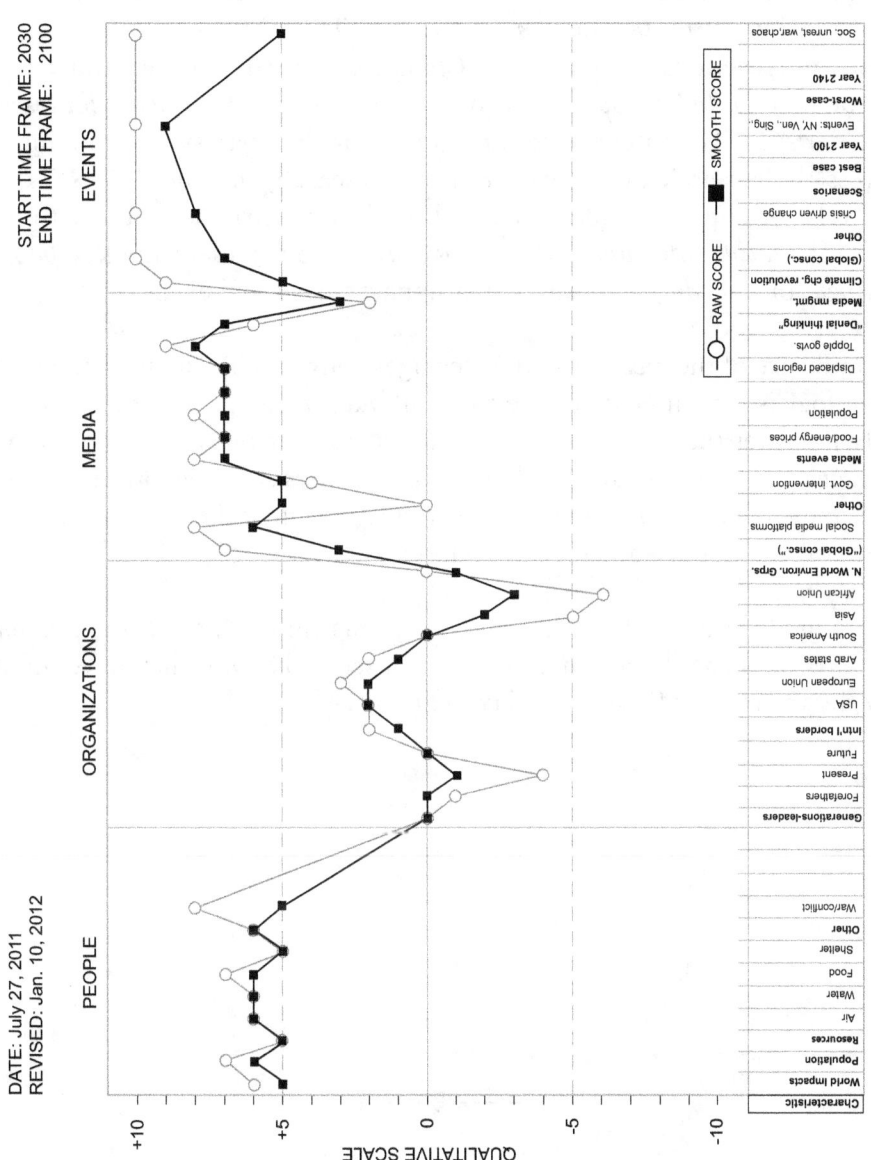

Chart 5.4 (rs): Climate Change: One Dramatic Moment Attribute -
Raw and Smooth Scores

Section 5.5: Climate Change Compilations

Parameters in this case study were outlined for the various *Climate Change* tipping point *attributes* and *characteristics*. The goal was to identify the actions and reactions involved in the *Climate Change living story* and identify the potential *global consciousness* inter-relationships of these *attributes* and *characteristics*. The intent here was to present the analysis and findings to address the research questions. From the research, the *characteristics* and their relationships were identified and used in construction of our tipping point *qualitative ratio analysis framework*. These represent the key *people, organizations, media, and events* that impact this global event.

The influence of the *media* and the messages delivered by and with the media technology is a complicated issue to identify a *future tipping point*. One can easily interpret these *attributes* and *characteristics* from many perspectives. However, the intent of this *qualitative ratio analysis framework* is to define *why?* and *when?* the *one dramatic event* might occur and compare this to the other two case studies for similar analysis.

The method for analyzing these *attributes* was simply look at the *sequence of events* in chronological manner, and observe the way they develop and how the various *media* sectors play their role.

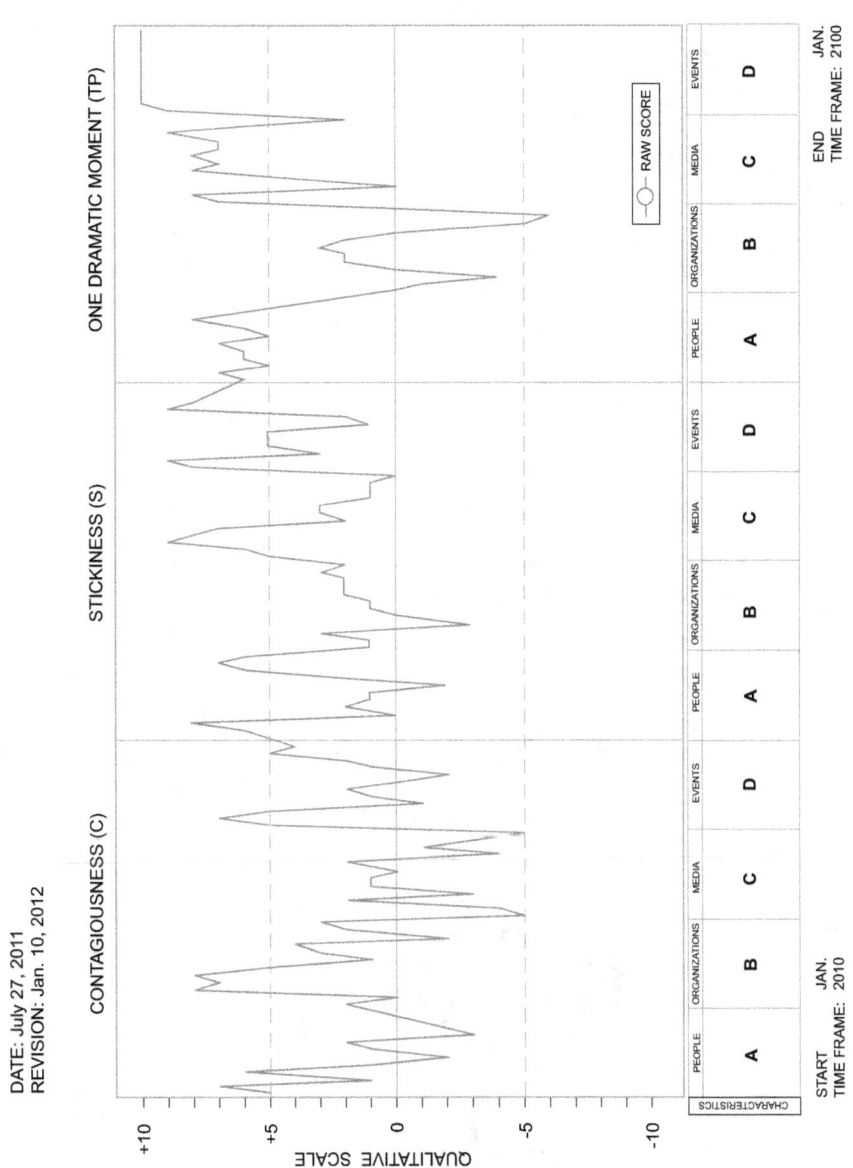

Chart 5.5 (r): Climate Change: Compilation: Raw Scores

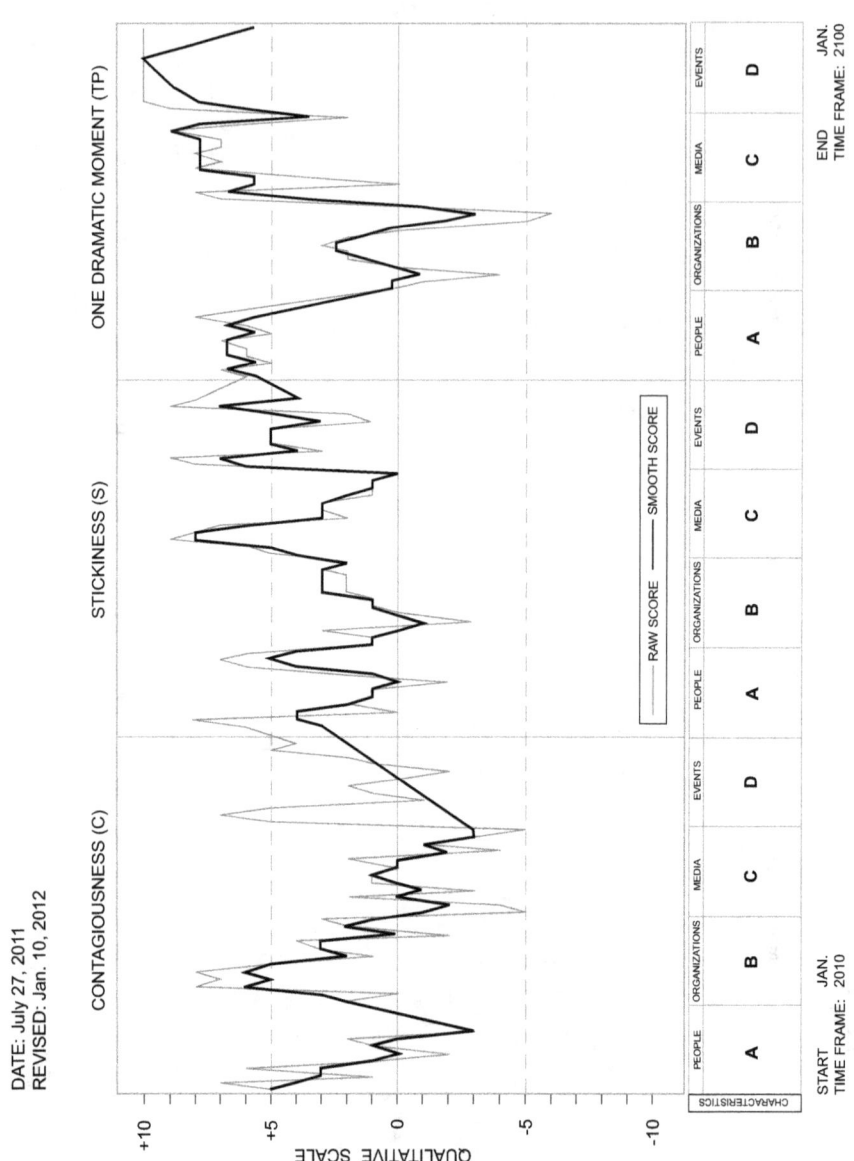

Chart 5.5 (rs): Climate Change: Compilation: Raw and Smooth Scores

Part III

Comparisons and Historical Context

CHAPTER 6: COMPARISONS
AND HISTORICAL CONTEXT

Section 6.1: Introduction

When comparing the *media impact* on *global events*, part of the criteria for the starting point was developing a representation for the appropriate qualitative measurement of the *attributes* and their *characteristics*. This was done in the preceding *Chapters Three, Four* and *Five* with the *qualitative ratio analysis frameworks* that highlighted and identified when the *tipping points* had (will) occur(ed).

The other important issue is that *media information* remains territorially bound in many respects, i.e. whereby it is limited by where one is able to find and analyze data. For example, much of the data and resources regarding *Climate Change* were found relative to the United States. In that regard, it underscores the view on what the global media culture might miss presenting key issues during an era in which *media* flows consistently over national borders and differing climate and cultural regions when it is presented different languages and different formats.

Based on these considerations, the preference leans towards the *trans-cultural approach* to the historical context and inevitably in the comparisons of the case studies. This approach assumes beginning with one state as a natural center of a comparison, in this case the US, but in effect, it points to more complex issues arising for carrying out the comparisons between different *media* sectors and the three case study *attributes* and *characteristics* that cross over territorial borders. To make these comparisons understandable, we must first agree that the research (largely detailed and published in *Media Tipping Points* (2012) has a bit of *container thinking* to its structure and how the research was presented.

Overall, this graphical comparison approach provokes us to think about *when* and *why tipping points* occur and their applications to *global events* in such a way that territoriality and associated politics, while important, were not wholly part of the legitimate perspective when considering the outcomes.

Traditional or *old media* is indeed a communication process that sometimes can be focused around relatively centralized power structures, in contrast to *new mass media*, which is marked by a more complicated, multi-centered power structure, for example, the Internet. In this context, the approach for the historical comparisons of the case studies is from a highly generalized perspective with the neither the intent to either validate or invalidate the *phenomenon* about *tipping points*, which were analyzed and presented earlier and the media environments presented in *Chapter Two*. What is problematic for the comparisons, particularly when considering the third case study, is that they can be criticized for the generalities that were made given the complexity of the size and magnitude and diversity and sheer quantity it encompasses, and, it has not ended in duration.

Assumptions from the beginning as to how to interpret agendas for national cultures and the application of the *tipping points* approach may be more harmful than helpful. As such, in this context specific cultural interrelationships are difficult at best to compare and outline herein. This permits us to focus on the *defined framework* of the *sequential attributes* and *characteristics* of our three case studies as a means to exhibit, compare and contrast the relationships with other historical *tipping point phenomena*, which may have more substance.

Having said that, the Internet is an effective medium for assembling and analyzing research. At this juncture in time, *media* has become a formidable power on the global stage with the rise of multimedia conglomerates, deregulated and private communications systems, and advanced technological innovations. With the globalization of *media* and *communication*, media technologies and industries have created a global culture in which people from all over the world can watch and share experiences, from war and humanitarian disasters, to sports and entertainment, to the free-market capitalism and advertisements, which make us all the world's media consumers.

Extracting from the analysis in the preceding chapters, we compare the *attributes sequentially* for each of the three case studies with the goal of comparing the *relationships* between them and a historical context. This comparison and the compilations are organized by the relative groupings of *people, organizations, media,* and the *events*.

These groupings identify and describe the *relative* influence. This study of the relationships evolves and become more important as we compare these in context with the *media* systems in place using the *attributes* themselves and as a benchmark for understanding some of the *media* developments and technological advances.

In any case, the third case study, *Climate Change*, poses the critical overall research question:

How to predict the one dramatic moment for the world and to understand (and anticipate) to the extent that these attributes evolve within complex global media systems, which can be engaged, manipulated and executed to influence the outcome?

Additionally, all of the *attributes* and *characteristics* that are highlighted in the case studies, ranging from a shorter timeframe of approximately one year for the *Obama Presidential Campaign*, to the duration of three years for the *International Financial Crisis*, and the on-going current and future events related to *Climate Change*, are similarly characterized by specific *people, organizations, media,* and *events* that have an influence.

The following offer the comparisons of these studies and a historical context, highlighting their similarities and differences and providing observations that can be derived from them. Obviously, *media* types are the focus for this exercise - their development, and influence on populations and the structures of economies. That focus then turns to the development of these *media* types with historical context for each of the case studies. An interesting aspect that emerges from the application of this exercise is that significant *global events* that occurred in the past followed similar *tipping points* evolutions (or *sequence*), which in effect led to their *one dramatic moment*, yet perhaps the means for analyzing and predicting them with current state-of-the-art methods is not available. In this context, the varied historical compilations that were generated with their *attributes* and *characteristics*, demonstrating a comparison of the case studies yields some striking similarities. However the respective *attribute*s reveal significant differences in these case studies, reflecting perhaps the technological sophistication and limitation of media types at the time for influencing *tipping points*.

The *media* issues that are highlighted in the case studies cover a range of examples that are sometimes extremely small, not only in absolute qualitative terms, but also relative to the size of audience of the media environments at the time. Despite their varying scale and magnitude, the case studies exhibit considerable *comparison value* when searching for similarities and differences that might be extrapolated towards even larger applications.

It ultimately rests with the world's nation states where specific questions pertaining to the *effect* and *impact* of these *phenomena* need to be addressed and where specific resources might be employed for implementing methods as a means of predicting and managing global *tipping points*, particularly *Climate Change*. The case studies vary widely, but when combined and compared, one can perceive certain common traits, most notably for *Climate Change* with regard to the influence between media on populations and response.

From a cross-section of each *attribute* and their *characteristics* that occurred within the case studies, it is interesting to note the interdependence of the factors as they are influenced by the *media* on these *global events*. This is not to say that all *global events* are capable of being analyzed and compared in this fashion, which is due to the complexities and size, and arguably, some experts will claim, as they have throughout much of history, that the very development and existence of such *phenomena* are associated with external factors outside human control. Rather, the comparisons show results of *modern media* forces acting out in the environments that affect smaller nation states as they are not independent enough on their own to avoid being part of the larger media *phenomena*.

One can conclude from these factors that the *comparison value* of the case studies is relatively high, albeit some questions remain relative to the strategic implementation and adoption to address such large content of issues, which will mandate all present and future nations and state involvement for finding solutions, again, particularly with *Climate Change*.

The following *comparisons* highlight the *media influence* during the *sequential attributes* for each of the case studies for the purposes of exhibiting where the similarities and differences are relevant, how these affected the outcome and are depicted in the *graphic summary* (using a basic *data visualization* technique) for all three. The intent is to illustrate the *attribute* and *characteristic* similarities and differences over the evolving *media lifecycle* of each *event* in order to draw some illustrative conclusions further to their existence and relationships. The patterns and *trends* that emerge via this synthesis of this application provide further discoveries about the nature of *global events* and the *media impact*.

Section 6.2: Tipping Point Attribute Compilations: Historical Comparisons

Putting the *Tipping Point attributes* and *characteristics* into a historical context provides challenges for exhibiting the influences of *media* over short and long period of times. Even with the evidence of the particular characteristic, some of these influences may be difficult to identify. However, for the *attribute* and *characteristics* identified in each of the case studies, we propose to also visit the measures of *similar events* and exhibit, on a relative scale, the *media impact* in terms of their tipping point *attributes* and *characteristics* as identified for each *global event*. The historical comparison of the three case studies highlight that there are *people, organizations, media, and events* inherently critical to development and potential for a tipping point to occur, and with basis for influence from the media sectors

The purpose is three-fold:

1. Contrasting and comparing the *attributes* and *characteristics* from a very historical context and broadly identify similarities and differences,

2. Identifying fundamental *trends*, media delivery systems and technology that may be prevalent, and,

3. Providing for the three case studies to be *measured qualitatively* relative to other *global events* with similar *characteristic* sequential importance and influence on the *tipping point attribute*.

It should be noted that there are limits to the number of similar *events* and quantity of *characteristics* identified for historical comparisons and the sizing the scope and magnitude accordingly for this version.

Using the *Tipping Point Methodology* as applied in the case studies certainly provides a basis for historical comparisons. *Media* may hold some very real options for how we as a civilization go about the business of the future *global events*. The approach strives to open up the opportunity to pursue additional questions that before had been effectively impossible to conceptualize.

For example:

Can we characterize the dynamics of media impact on key global events?

Does information change and influence the evolution of the tipping point baseline as it evolves and propagates?

Is it be possible to use this tipping point model over long periods of time, in a way that is essential to discovering the core solution to other world event and ultimately, climate change?

Granted, one could combine the approaches here into several multidiscipline approaches towards addressing scientific, political, cultural and financial issues associated with this *phenomenon*. And more generally, it appears useful to further understand the *media impact* on the different *tipping point attributes* and *characteristics* as exhibited during *global events*.

Extracting from the detail in the proceeding chapters, we present the case studies with the goal of understanding the *attributes, characteristics* and relationships in a historical perspective. Looking at the case studies research questions that were posed earlier, one can see how the *attributes* took place in relative groupings of *people, organizations, media, and events,* and the net results. We attempt here to contrast and compare, identify and/or qualify the relative impact among them with similar historical *events*. This study of the relationships evolves and becomes more important relative to the fundamental research question of trying to point out that by analyzing these *attributes* it might be potentially possible to influence and predict and/or help avoid a potential *global tipping point,* in the case of *Climate Change,* from occurring.

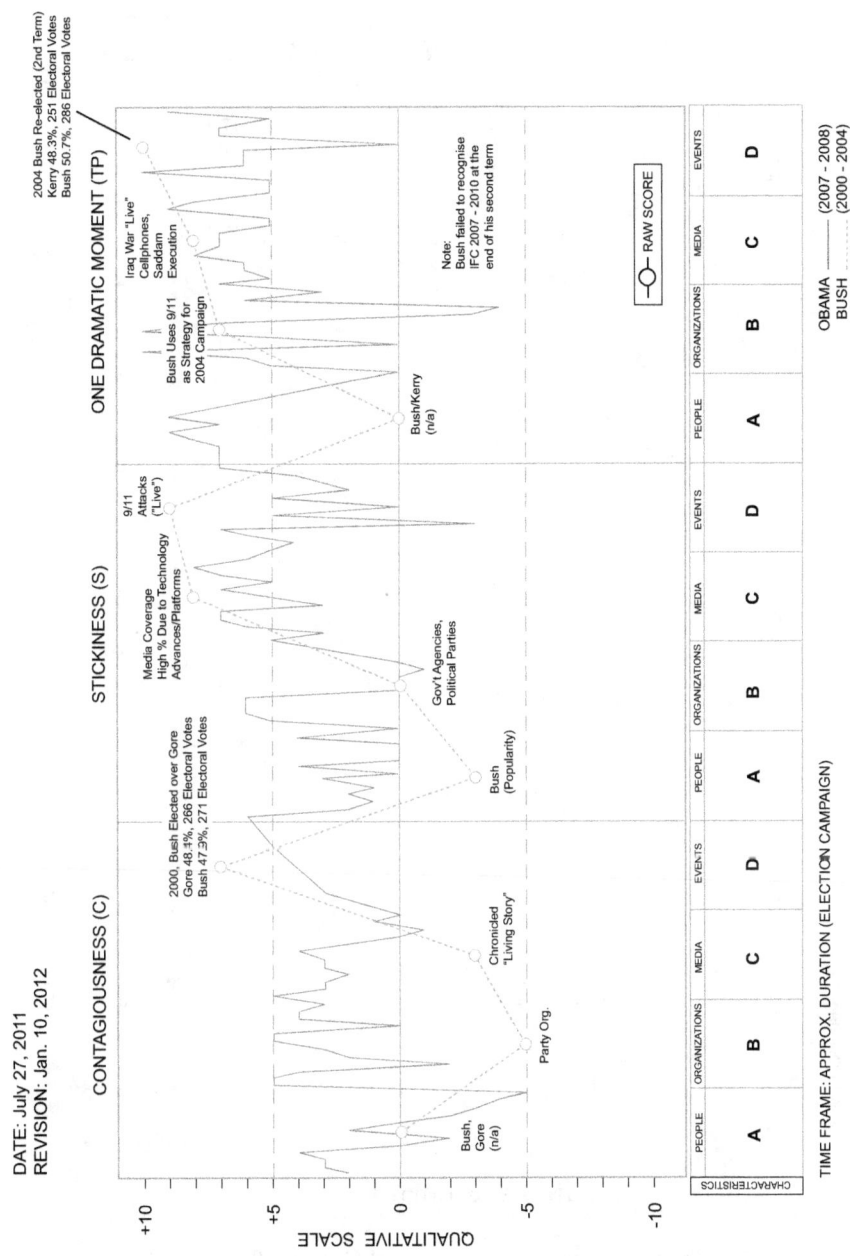

Chart 6.2: Obama Presidential Campaign (2007–2008)
Historical Comparison

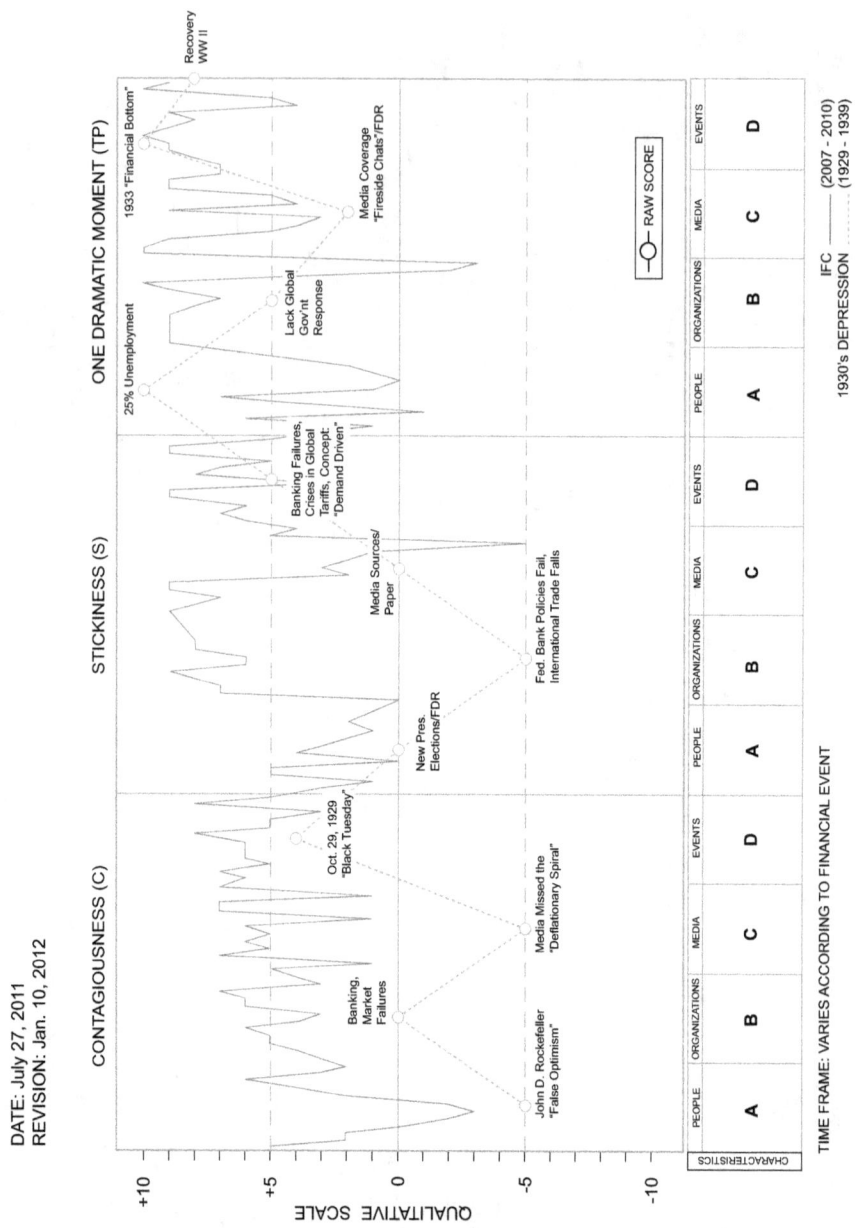

Chart 6.2: International Financial Crisis (2007–2011)
Historical Comparison

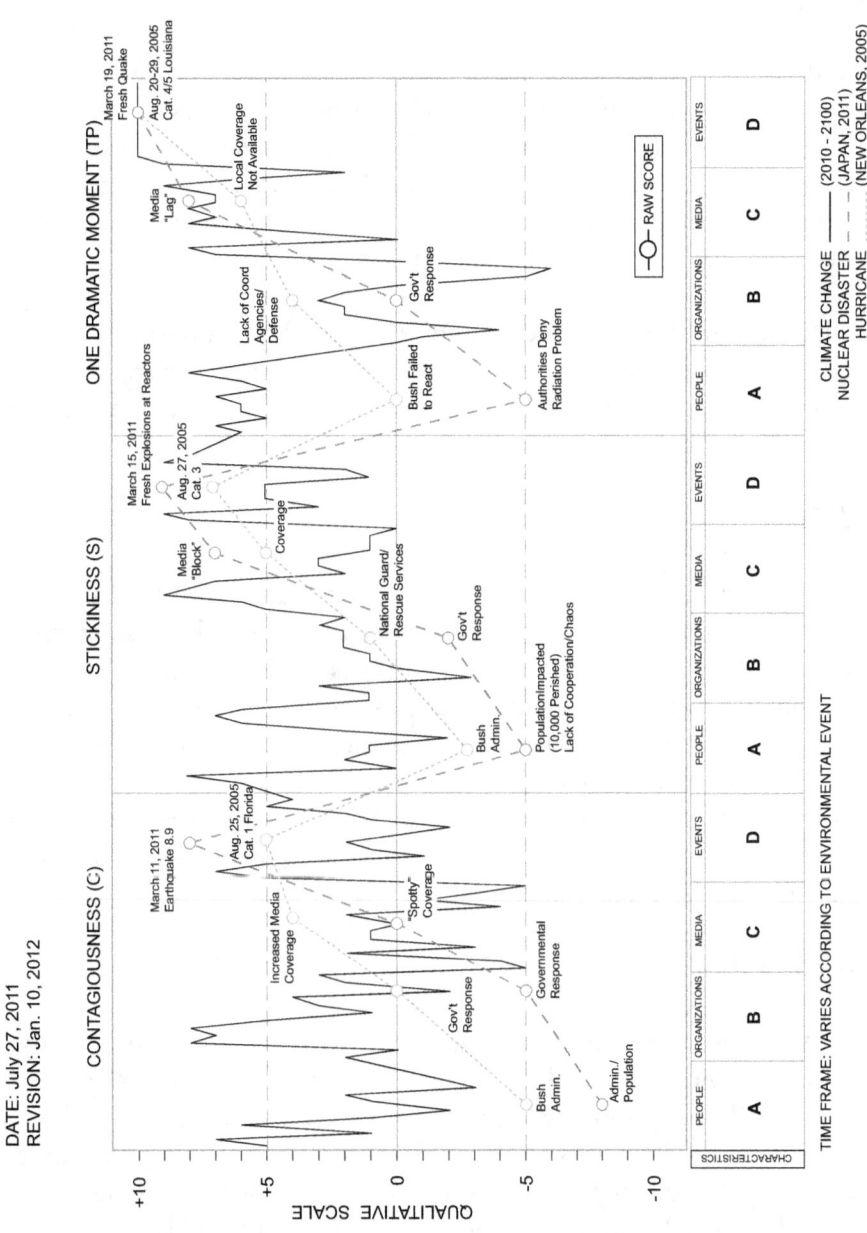

Chart 6.2: Climate Change (2000–2030) Historical Comparison

Section 6.3: Media Development: Global Events: Historical Dateline

A few decades ago *media* reporting was through newspapers, radio, and television. Things are different now (2013) as we are witnessing a revolution of people-oriented reporting of *global events*. This element of intimate knowledge of each *event* being reported has dramatically changed. This revolution has intrinsically altered the way *global events* become more personal and more accurate – however more subjective.

As highlighted in the case studies and analysis and comparisons, there are three (3) key media development *milestones* when comparing *historical media development* relative to a *"dateline"* context (calendar periods) that can now be best described as:

"Real Time"

The *real time cyber culture* is a *phenomena* associated with the Internet and network communications and are the *cultural objects* that use technology for distribution and exhibition. *Media* today is a mix between older cultural conventions for data representation, access, (and manipulation), and recent technological conventions. In order to incorporate this milestone and understand its maximum benefit, development of a much more comprehensive comparative analysis would be required, which would correlate the history of technology with social, political, and economical histories with the significant sequential deployments.

"Living Story"

Living Story is the *avant-garde* new way of accessing and analyzing information, whereby, *algorithms*, are essential part of the *living stories. Living Stories* depend on technology, and can be driven and executed by humans. These morphic *media* platforms have increased the quality and accuracy of the content and frequency of communication, whereby influencing outcomes and providing insights about and between people all over the world.

"Global Consciousness"

Changes in the *new media environment* have (and will) create a series of key milestone with the concept of *global consciousness* influence. The *trend* of *global consciousness* is not only an evolving geographically, expanding from a nation to worldwide, particularly as the population of the world increases and barriers to entry and access are removed. *Virtual communities are being established online and transcend geographical boundaries, thereby eliminating social limitations. This *new media* connects like-minded others worldwide and provides a foundation for increasing *global consciousness.* Society doesn't follow a script based on the course of technological change, since many factors, including individual inventiveness and entrepreneurialism, (and conflict) intervene in the process of science, technical innovation and social interactions.

The following graphic: *Dateline of historical Tipping Point Milestones,* applies time and population growth in a *media development and scale of influence* context.

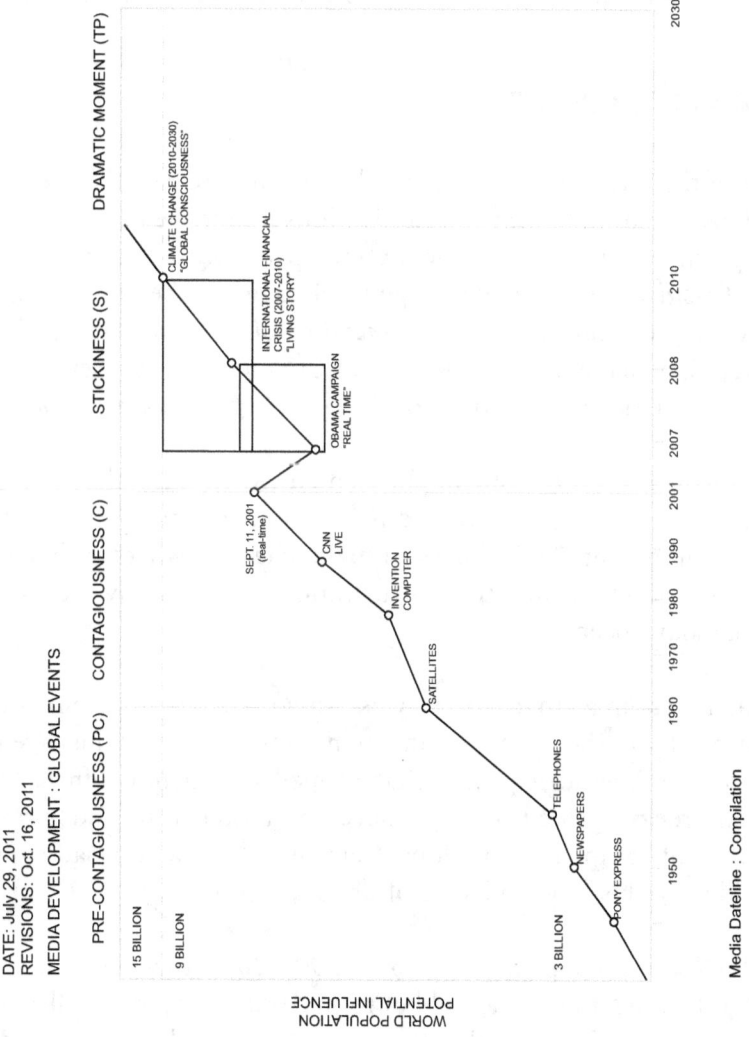

Fig 6.3: Media Dateline Compilation

Placed in a historical perspective, the question remains, whether, in fact, this historical perspective supports the leading edge equation for understanding the current and future context for *global events*.

Ultimately, these milestones show what has driven world societies closer together (or further apart).

Would similar tipping point attributes and characteristics be evident if one were to expand the scope and complexity of this version?

Would this support the notion (or not) the great shift taking place in the world now, proving to be less about the domain of individual cultures and out-of-date conventions and more about global consciousness?

Section 6.4: Summary

For each of the three case studies, one clearly could review the *media content* and make some broad-stroke generalizations about the occurrence of these *tipping point phenomena*. However, the *spreading contagion* during the *Obama Presidential Campaign* piggybacked with the plummeting stock market during the *International Financial Crisis* and the current dazed and confused policy makers facing the issue of *Climate Change*, merely reinforces the basic premises and evolutions of the *tipping point phenomena*.

As each of these *tipping points* approach(ed), with unthinkable and finally irreversible end results, it prompted populations to take radical action and required governments and leaders to make changes whether they liked the outcome or not, sometimes to the frustration of many, however drastic the moves may have been.

One might assume that there was some carryover from the Bush administration to the Obama campaign that reinforced his agenda and pledges to the American people that propelled him into office. However, the *media* frenzy created a huge source of momentum and ultimately a huge source of campaign donations. Certainly there was broad support for McCain during this time and a lot of unhappy voters.

During the *International Financial Crisis 2007-2010*, there was the consistent failure by the *media* (and the public) to understand conceptually the broad spectrum and complexity of the crisis that was unfolding. How widespread and serious the crisis was, for example. At the time (and maybe even now), we don't know. Many months have passed since this *living story* was carried. Headlines show the nation states throughout the world are still suffering from severe financial and budget troubles. Governments lack the resources for a full investigation and a means for correcting the magnitude of the problems.

With the *International Financial Crisis 2007–2010*, the banking industry had little incentive to correct their processes to lead to their insolvency that deconstructed the markets in late 2008 and early 2009. But those markets have long since returned to normal, in part because everyone knows that banks will be bailed out if they get in trouble, as was the case throughout the European Union in mid 2011–12, with the issues surrounding the Greek, Italian, and Portuguese economies.

The fact is with many of these stories you can't believe what is being told in the *media* and who is accountable. As US officials are grappling with increasing the national debt levels reaching to gargantuan proportions, one wonders where it all is leading, as the average American taxpayer each (in August 2011), owed 125,000 USD as part of the national debt.

Who can you believe? Recently (early July 2011), a *global event* within the *media sector* exhibited similar *tipping point phenomena. The News Corp.* (UK) phone hacking scandal that involved the Murdoch publishing empire, the *British Government* and *Scotland Yard* played out in macro fashion (in billions of dollars in value, lost revenue, and loss of life) with the same *attributes* and *characteristics* of the prototypical tipping point case study.

Within a period of weeks the *world media audience* was riveted on the *contagiousness, stickiness*, and what lead to a *dramatic tipping point* for the Murdoch Empire relative to the scandal that unfolded crossing economic, cultural, and political lines within British society and media environments. The *characteristics* were easily identified in a tipping point context of *people*- -the editors, the police commissioner, reporters, the politicians, and the presidents of companies; the *organizations*—news organizations, political parties, parliament, and the Metropolitan Police Department; *media*—both as a participant–*News Corp.* and the watchdog journalists that covered the story for the various other media sectors; and *events*–illegal hacking of telephones, e-mails and voice mails, payments and settlements to avoid scrutiny, the evidence that was ignored, the players that were arrested, hearings that were held in broadcast live, the percentage of news coverage by *media* in covering a story of one of their own, the government and corporate apologies that could be analyzed much in the same way as we have done with our case studies with the qualitative ratio analysis framework which would show when the tipping point occurred in this recent and somewhat ironic *global event*.

This *event* did nothing more than compete for the *media space* in a world that has now lost objective perspective on the information that is being presented to them. Remembering, in comparison to scientific *phenomena* that is analyzed and reviewed by peers, the *real-time* unfolding of this *event* had no *fact checking* or means to qualify the information that was being presented. Most individuals did not have the filters, or the energy, or the expertise to evaluate and analyze the substance in the reporting and the relevance of the topic.

This was an easier *media* story to tell. As was the 2012 *Obama Presidential Campaign* which was determined by its *one dramatic moment*: Hurricane Sandy.

All the while, the world's most pressing issue of *Climate Change* has moved to the back pages of the *global consciousness*. Our analysis here underscores the reality that the clock is ticking, and we have no time to lose.

In January 2013, during his inaugural speech for his second term as president of the US, Barack Obama placed *Climate Change* at the top of his priorities for the coming four years. Was this a genuine pledge of commitment to action, or, a seasoned *media persuader* piggybacking a trending *living story* topic for political gain? Obama is no fool.

Part IV

Summary

CHAPTER 7: CONCLUSIONS

The application of the *Tipping Point Theory* in the context of *global events* and *media impact* provides a *conceptual framework* for understanding the role of *media* and international *events* and provides for understanding of how society can respond and develop around it. Throughout this book, the *tipping point attributes* and *characteristics* are recognized as a means by which international *events* may be analyzed in the context of *media* influence.

Underlining these *phenomena* is the notion that *events* on a worldwide scale can be analyzed and predicted. A simple application of this *phenomenon* suggests that if a window is broken and left unrepaired, i.e. the international financial system, whereby, if it appears that the system is unregulated, unmanaged, or that no one cares about it and no one is in charge, soon more windows will be broken, and the condition will *spread*. In countries or regions with lack of financial oversights and monitoring regulation is all the equivalent of *broken windows*.[112] Perhaps, in keeping with the example, such *broken windows* promote invitations to further financial *tipping points* - as is evident in events in Ireland and Greece that affected the entire European Union in 2011–12.

This is the *epidemic* of finance. The effects are *contagious* – just as a *trend* is *contagious*. It can start with just one broken window (or country financial failure) and spread like a *virus*, sometimes instantaneously, particularly if the message is delivered via television or internet, to an entire region or population in the world. The impetus to engage in a certain type of behavior is not necessarily coming from a certain kind of person or culture, but from a feature of the environment, i.e. the financial sector–banking.

Further, the *Broken Window Theory* postulates that if a particular behavior in a community (or world) goes unaddressed, it signals that nobody cares about the community (or world) resulting in additional behavior of the same type. Given that current information technology has evolved with the

---

112 Wilson and Kelling argue that impacts and events are the inevitable result of disorder.

Internet, text messaging, *Twitter*, and a *24-hour media cycle*, this creates a combustible mix indeed. This also provides an opportunity for the application of our *conceptual framework* for understanding *tipping points* to evolve together with the idea of embracing *international momentum* or the *global media consciousness* to create legal instruments and policies with these developments that might influence the world's most pressing issue*: Climate Change.*

In addition, as a potential mechanism for international relations, *media tipping points* and *global events* has a wide variety of applications; among them is a process of dissemination of unbiased, factual information and resolution of disputes between countries and regions. Since some countries and regions are driven by different growth and economic objectives, particularly between developed and undeveloped nations, understanding the context of their potential tipping point contributions, actions or reactions and how this is being portrayed by the *media* and disseminated on a worldwide scale.

"*What if the pollution coming from our island nations was threatening the very existence of the major emitters?*" stated President Marcus Stephens of the tiny island state of *Nauru*, speaking on behalf of some 14 island states vulnerable to disappearing. "*What would be the nature of today's (climate change) debate under those circumstances?*"..."*Climate change is a threat to international peace and security,*" Mr. Stephens said to the *UN Security Council* in June of 2011, comparing it to nuclear proliferation or terrorism given its potential to destabilize governments and create conflict.[113]

The application of the *Tipping Point Theory* has the potential to change the very nature of our approach to these *phenomena* as societies start to understand the interdependent relationships and responsibilities. The *Tipping Point's conceptual framework* can offer an alternative path towards analyzing and adjusting future impacts. The cases examined in this current version demonstrate that *tipping points* analyzed in the context of *media* and its impact on *global events* shows, indeed, that there can be success in viewing the *trends* and *sequential* development of world events in this manner. It could be also argued that for this reason *Tipping Point Theory* could be used to address the media role with regard to the *Climate Change* issue.

It is recognized that the *conceptual framework* that has been put forth is an option which may not fully explain all the results beyond the initial premises. While there may be several possible explanations for this, one,

---

113 "The Environment in the News," United Nations Environment Programme, July 22, 2011. Retrieved from www. unep.org/cpi/briefs/2011July22.doc, p. 9.

is the particularly difficult task of trying to analyze and assemble all of the data, opinions, and research relative to the size of each specific case study. As evidenced by the data and documentation employed in presenting this framework, there are virtually limitless opportunities to refine and focus the strengths of these arguments.

Having said that, the fact that *Climate Change* is such a large-scale *global event*, the application of *Tipping Point Theory* in this context may yet to give birth to a known practice and a further refined systematic process for recognition, analysis, or consideration as a potential model for addressing the media impact on *global events*, even though this potential was recognized in the isolated comments and sources that were found and cited throughout the three case studies.

Stated another way, the *Tipping Point Theory* approach to *global events* can become incorporated into the structure of the international system of *media* relationships and could become almost routine for countries to incorporate the same *global consciousness* objectives - such as establishing a secure *Climate Change* agenda with deployment of resources and technological means coupled with diplomatic missions and infrastructure as a way of meeting the goals associated with such a formidable task.

Despite the argument that *tipping points* are connected to *global events*, which unfold and are influenced during the *attributes* and *characteristics* in the analysis here, it is likely that a parallel future objective to addressing the tipping point for *Climate Change* should recognize and focus on the lack of knowledge and the understanding of the impacts by the world's growing population on our environment and how *media* holds the key to our very existence.

The degree of certainty that it is necessary for our world leaders to develop solutions merely serves the dispute that this, as with any large-scale human endeavor, requires the deployment and execution of new concepts and approaches mandated by survival on the world stage.

A committee panel of scientists, former government officials, and national security experts in October of 2011 were already urging the US government to begin researching a radical and drastic fix to avoid *Climate Change* disaster by directly manipulating the Earth's climate to lower temperatures. This committee proposed extreme engineering techniques, which included scattering particles in the air to mimic the cooling effect of volcanoes and stationing orbiting mirrors in space to reflect sunlight. The panel stated that it was time to begin researching and testing such ideas, "...*just in case the climate system reaches a tipping point and the swift remedial action was required.*"[114]

Everyone has to understand what is at stake. Basically the current body of *Climate Change* proposals fall into two broad groups, one, which is widely known as *climate remediation*, which concerns carbon dioxide *removal* from the atmosphere. What is good about these is they are "*generally uncontroversial and don't introduce new global risks,*" according to Ken Caldeira, a climate expert at *Stanford University* and a panel member.[115]

The controversy arises more with the second group of techniques, termed *solar radiation management* approaches, which involve physically manipulating the amount of solar energy that bounces back into space before it can be absorbed by the year. The problem with these techniques is that they pose a risk of upsetting the Earth's natural rhythms. With them, Dr. Caldeira states, "*The real question is what are the unknowns: Are you creating more risk than you are alleviating?*"[116] Joe Romm, a fellow at the Center of American Progress, has made a similar point, comparing geo-engineering to a dangerous course of chemotherapy and radiation to treat a condition that otherwise is curable through diet and exercise or, in this case, emissions reduction.

The bottom line is that many countries fault developed nations and the United States in particular, for government inaction on *climate change*, especially given its longtime role as the chief contributor to the problem. Frank Loy, a former Under Secretary of State who is now the nation's chief climate negotiator, suggests that people around the world would see past these issues if the United States embraced geo-engineering studies, provided that it was "*very clear about what kind of research is undertaken and what the safeguards are.*"[117] Clearly media could deliver these messages.

The case studies in this version showed that there is nothing inherent in the *phenomena* of the *tipping point attributes* and *characteristics* that would prevent such an application from being used in resolving the issue of *Climate Change*. A broad range of factors involved with *Climate Change*, the history and future development of *media* and *global events* suggests that we stand a better chance of yielding positive results in some circumstances than others.

---

114 Cornelia DEAN, "Group Urges Research Into Aggressive Efforts to Fight Climate Change," *New York Times*, October 4. 2011. Retrieved from http://www.nytimes.com/2011/10/04/science/earth/04climate.html.
115 Ibid.
116 Ibid.
117 Ibid.

The *tipping points* analyzed in the *Obama Presidential Campaign* and the *International Financial Crisis 2007–2010,* viewed by measurable impact, point out the difference required for the issue as large as *Climate Change.* Whereby, there were many *attributes* and *characteristics* shared by first two case studies that can be applied to *Climate Change,* and in a sense, the *conceptual framework* produced a positive result, *vis-à-vis* a methodology for categorizing and exhibiting the specific *attributes* of *contagiousness, stickiness,* and the *one dramatic point* in their *sequential* evolution and the media's role and importance.

From another perspective, the analysis utilizing *Tipping Points Theory* would appear to at least offer a chance of success in addressing the *media impact* on *Climate Change* issue, given the scope, magnitude and impact of the event in terms of population effects, country resources, and/or economic activity involved, which if not resolved, could become the subject of a very high intensity conflict worldwide between nations. Meanwhile, *attributes* that are unique to the *Climate Change* issue must be taken into account when weighing the viability of using *tipping points* and evolving *media* technology to address them. Thus, *tipping points* as applied in this context to *media impact* and *global events* may be a means of addressing certain *global events* such as *Climate Change.*

This also raises the question as to whether or not applying *Tipping Point Theory* universally to *media impacts* on all *global events* is legitimate. This remains a subject for further research. Nonetheless, the case studies presented here fall within the limits of what can be analyzed, and what can be used conceptually for addressing the further developments of the *Climate Change* issue and other *global events* with similar characteristics. Further applications of this concept in the context of *media impact* and *global events* may bring to light more clearly where the opportunities are and allow a more refined approach to be developed. The case studies presented here provide evidence that the *phenomenon* of *tipping points* as a means of analyzing *media impact* and *global events* is certainly practical on an international scale, and with greater visibility of this approach and practice, would bring it in more fully into focus with acceptance as a recognized *conceptual framework.*

Media Impacts and Global Events Website:

For further information please visit the: *Media impact and Global Events Website*:

http://groupemdg.typepad.com/global_media_and_world_ev/

This website tracks, highlights, categorizes and provides articles and commentary related to Tipping Point *Attributes and Characteristics*, as well as information and recent updates related specifically to the three case studies included in this version and other media and global event issues.

Bibliography

Part I: References

Chapter 1: The Tipping Point Theory

CONDON, William, "Cultural Microrhythms," in *Interaction Rhythms*, 1974, New York, Human Sciences.

FRIEDMAN, Thomas, *The World is Flat: A Brief History of the Twenty-first Century*, New York, Farrar, Straus and Giroux, 2005.

GLADWELL, Malcolm, *The Tipping Point: How Little Things can Make a Big Difference*. New York, Back Bay Books, 2002.

GRANOVETTER, Mark, "Threshold Model of Collective Behavior," *The American Journal of Sociology*, 83, 1978, pp. 1420–1443.

GRODZINS, Morton, *The Metropolitan Area as a Racial Problem*. Pittsburgh, University of Pittsburgh Press, 1958.

LINKOLA, Pentti, "The Doctrine of Survival and Doctor Ethics," 1992.

National Security Agency (NSA), *Information Operations Roadmap* (National Security Archive Electronic Briefing Book No. 177), October 30, 2003.

POSTON, Tim and STEWART, Ian. Catastrophe: Theory and Its Applications. New York: Dover, 1998

SCHELLING, Thomas, "Dynamic Models of Segregation," *Journal of Mathematical Sociology*, 1, 1972, pp. 143–186.

United Nations Department of Economic and Social Affairs/Population Division (UNDES), "World Population in 2300 to be around Nine Billion Persons," December 9, 2003. Retrieved from http://www. un.org/apps/news/story.asp?NewsID=12439

U.S. Census Bureau, *International Database, June 2011 Update, World Population 1950–2050*, 2010. Retrieved from http://www.census. gov/population/international/data/idb/worldpopgraph.php

WELLS, Gary and PETTY, Richard, "The Effects of Head Movement on Persuasion: Compatibility and Incompatibility of Responses," *Basic and Applied Social Psychology*, 1, 1989, pp. 219–230.

WILSON, James and KELLING, George, "Broken Windows. The Police and Neighbourhood Safety," *Atlantic Magazine*, 3, 1982, 29–38. Retrieved from http://www.theatlantic.com/magazine/ archive/1982/03/broken-windows/4465/

Chapter 2: Global Media

ATTON, Chris, "Reshaping Social Movement Media for a New Millennium," *Social Movement Studies*, 2, 2003, pp. 3–15.

BOYD, Clark, "BBC News – North Korea creates Twitter and YouTube presence," BBC, August 18, 2010. Retrieved from http://www.bbc. co.uk/news/world-us-canada-11007825

BRODZINSKY, Sibylla, "Facebook Used to Target Colombia's FARC with Global Rally," *The Christian Science Monitor (Boston)*, February 4, 2008. Retrieved from http://www.csmonitor.com/World/ Americas/2008/0204/p04s02-woam.html

CARVIN, Andy, "Welcome to the Twitterverse," *National Public Radio*, February 28, 2009. Retrieved from http://www.npr.org/templates/ story/story.php?storyId=101265831

CASTELLS, Manuel, The Rise of the Network Society (The Information Age: Economy, Society and Culture, Volume 1), Hoboken, Wiley-Blackwell, 1996.

CLEE, Nicholas, "The Bookseller," *Guardian Unlimited*, March 1, 2003, Retrieved from http://www.guardian.co.uk/books/2003/mar/01/featuresreviews.guardianreview30

CROSBIE, Vin, "What is New Media?" 1998, *Sociology Central*. Retrieved from http://www.sociology.org.uk/as4mm3a.doc

CROTEAU, David and HOYNES, William, (2003). *Media Society: Industries, Images and Audiences* (3rd edition). Thousand Oaks, Pine Forge Press, 2003.

DOUGLAS, Nick, "Twitter Blows Up at SXSW Conference," *Gawker*, March 12, 2007. Retrieved from http://gawker.com/243634/twitter-blows-up-at-sxsw-conference

DURHAM, Meenakshi, and KELLNER, Douglas, *Media and Cultural Studies: Keyworks (KeyWorks in Cultural Studies)*, Malden, MA and Oxford, UK, Blackwell Publishing, 2001.

EDELMAN, *Edelman Trust Barometer 2008*, 2008. Retrieved from http://www.edelman.com/trust/2008/TrustBarometer08_FINAL.pdf

EDELMAN, *Edelman Trust Barometer 2010*. 2010. Retrieved from http://www.edelman.com/trust/2010/

FELDMAN, Tony, *An Introduction to Digital Media*, London, UK, Routledge, 1997.

FLEW, Terry and HUMPHREY, Sal, "Games: Technology, Industry, Culture," in *New Media: An Introduction* (second edition), 2005, South Melbourne, Australia, Oxford University Press, pp. 101–114.

GOLDMAN, Russell (January 5, 2007). "Facebook Gives Snapshot of Voter Sentiment." ABC News. Retrieved March 23, 2010.

GRIER, Thom, *et al.*, "The 100 Greatest Movies, TV Shows, Albums, Books, Characters, Scenes, Episodes, Songs, Dresses, Music Videos, and Trends that Entertained Us Over the 10 Years," *Entertainment Weekly*, 1079/1080, December 11, 2009, pp. 74–84.

JACOBS, Andrew, "Chinese Woman Imprisoned for Twitter Message," *New York Times*, November 18, 2010. Retrieved from http://www.nytimes.com/2010/11/19/world/asia/19beijing.html

JOSHI, Rajmohan, Encyclopaedia of Journalism and Mass Communication: Media and Mass Communication, New Delhi, India, Gyan Publishing House, 2006, p. 95.

KAPLAN, Andreas and HAENLEIN, Michael, "Users of the World, Unite! The Challenges and Opportunities of Social Media," *Business Horizons*, 53, 2010, pp. 59–68.

KAZENIAC, Andy, "Social Networks: Facebook Takes Over Top Spot, Twitter Climbs," *Compete Pulse blog*, February 9, 2009. Retrieved from http://blog.compete.com/2009/02/09/facebook-myspace-twitter-social-network/

KELLNER, Douglas, "New Technologies, Technocities, and the Prospects for Democratization," in *Technocities*, 1999, London, UK, Sage, pp. 186–204.

KELLNER, Douglas, "Globalization and Technopolitics," in *The Future of Revolutions: Rethinking Radical Change in the Age of Globalization*, 2003, New York, Zed Books, pp. 180–194.

LEARY, Brent, "Overemphasis on Brand Building Leads to Mistrust, Inc., March 24, 2010. Retrieved from http://technology.inc.com/2010/03/22/overemphasis-on-brand-building-leads-to-mistrust/

LERNER, David, The Passing of Traditional Society: Modernizing the Middle East, New York, The Free Press, 1958.

LEVY, Steven, "Twitter: Is Brevity the Next Big Thing?" *Newsweek*, April 30, 2007. Retrieved from http://www.msnbc.msn.com/id/17888481/site/newsweek/

LINNELL, Nathan, "Social Media Influence on Consumer Behavior," *Search Engine Watch*, May 3, 2010. Retrieved from http://searchenginewatch.com/article/2049190/Social-Media-Influence-on-Consumer-Behavior

LISTER, Martin, DOVEY, Jon, GIDDINGS, Seth, GRANT, Ian, and
KELLY, Kieran, *New Media: A Critical Introduction*, London, UK,
Routledge, 2003.

MARMURA, Stephen, "A Net Advantage? The Internet, Grassroots
Activism and American Middle-Eastern policy," *New Media &
Society*, 10, 2008, pp. 247–271.

MANOVICH, Lev, "New Media from Borges to HTML," in *The New
Media Reader*, 2003, Cambridge, The MIT Press, pp. 16–23.

MCCOMBS, Maxwell and SHAW, Donald, "The Agenda-Setting Function
of Mass Media," *Public Opinion Quarterly*, 36, 1972,
pp. 176–187.

MCLUHAN, Marshall, *The Gutenberg Galaxy: The Making of
Typographic Man*, London, UK, Routledge and Kegan Paul, 1962;
Marshall MCLUHAN, *Understanding Media: The Extensions of
Man*, Toronto, Canada, McGraw-Hill, 1964.

MCLUHAN, Marshall and QUENTIN, Fiore, *The Medium is the Message*,
Hardwired, San Francisco, 1967, pp. 8–9, 26–41.

Press release, "Media Advisory M10-012 – NASA Extends the World Wide
Web Out into Space," NASA, January 22, 2010. Retrieved from
http://www.nasa.gov/home/hqnews/2010/jan/HQ_M10-012_ISS_
Web.html

MOROZOV, Eugenie, "The Net Delusion: The Dark Side of Internet
Freedom," *PublicAffairs*, 2011.

PRESTON, Paschal, Reshaping Communications: Technology, Information
and Social Change, London, UK, Sage, 2001.

Quantcast, "*Facebook*.com," October 31, 2011. Retrieved from http://
www.quantcast.com/facebook.com

REED, Thomas Vernon, "Will the Revolution be Cybercast?: New Media,
the Battle of Seattle, and Global Justice," in The Art of Protest:
Culture and Activism from the Civil Rights Movement to the
Streets of Seattle, 2005, Minneapolis, University of Minnesota
Press, pp. 240–285.

RHEINGOLD, Howard, The Virtual Community: Homesteading on the Electronic Frontier, Cambridge, The MIT Press, 2000.

ROBERTS, Laura, "North Korea Joins Facebook," *The Daily Telegraph (London)*, August 21, 2010. Retrieved from http://www.telegraph.co.uk/technology/facebook/7957222/North-Korea-joins-Facebook.html

SCHORR, Angela, SCHENK, Michael, and CAMPBELL, William, *Communication Research and Media Science in Europe*, Berlin, Germany, Mouton de Gruyter, 2003.

Second Life, http://secondlife.com/whatis/#Be_Creative

SHANE, Scott, Dismantling Utopia: How Information Ended the Soviet Union, Chicago, Ivan R. Dee, 1994.

SHANE, Scott, "Spotlight Again Falls on Web Tools and Change," *New York Times*, January 29, 2011, para. 12.

SNIDERMAN, Zachary, "North Korea's Newly Launched Twitter Account Banned by South Korea," *Mashable.com*, August 19, 2010. Retrieved from http://mashable.com/2010/08/19/north-korea-twitter-banned/

SULLIVAN, Michelle, "'Facebook Effect' Mobilizes Youth Vote," *CBS News*, November 3, 2008. Retrieved from http://www.cbsnews.com/stories/2008/11/04/politics/uwire/main4568563.shtml

TURKLE, Sherry, "Who am We?" *Wired*, 4.01, January 1996.

VOLKMER, Ingrid, *News in the Global Sphere. A Study of CNN and its Impact on Global Communication.* Luton, UK, University of Luton Press, 1999.

WARDRIP-FRUIN, Noah and MONTFORT, Nick (eds.), *The New Media Reader.* Cambridge, The MIT Press, 2003.

WASSERMAN, Herman, "Is a New Worldwide Web Possible? An Explorative Comparison of the Use of ICTs by Two South African Social Movements," *African Studies Review*, 50, 2007, pp. 109–131.

WELLS, Roy, "41.6% of the U.S. Population has a Facebook account," *Social Media Today*, August 8, 2010. Retrieved from http://socialmediatoday.com/index.php?q=roywells1/158020/416-us-population-has-facebook-account

WESSEL, Rhea, "Activist Investors Turn to Social Media to Enlist Support," *The New York Times DealBook*, March 24, 2011.

WILLIAMS, Raymond, *Television: Technology and Cultural Form*, London, UK, Routledge, 1974.

PART II: REFERENCES

Introdution and Methodology

BACHOR, Daniel G., "Rethinking Case Study Research Methodology," paper presented at the Special Education National Research Forum, Helsinki, May 2000; DAVIS, T. M. and BACHOR, Daniel G. "Case Studies as a Research Tool in Evaluating Student Achievement," paper presented at the Canadian Society for Studies in Education Conference, Sherbrooke, Canada, June 1999.

KVALE, Steiner. Interviews. An Introduction to Qualitative Research Interviewing, Thousand Oaks, CA, Sage, 1996; STRAUSS, Anselm L. Qualitative Analysis for Social Scientists, Cambridge, UK, Cambridge University Press, 1987.

LINCOLN, Yvonne S. and GUBA, Egon G., Naturalistic Inquiry, Beverly Hills, CA, Sage, 1985.

MUHR, Thomas, Atlas/ti: The Knowledge Workbench, Version 4.1, Berlin, Scientific SoftwareDevelopment, 1997.

STRAUSS, Anselm L. and CORBIN, Juliet M., Basics of Qualitative Research: Grounded Theory Procedures and Techniques, Newbury Park, CA, Sage, 1990.

VAILLANT, George E., Adaptations to Life, Boston, MA, Little Brown, 1977.

YIN, Robert K., Case Study Research: Design and Methods (2nd ed.) Newbury Park, CA, Sage, 1994.

Chapter 4: International Financial Crisis 2007–2010

ALTMAN, Robert C., "The Great Crash, 2008, A Geopolitical Setback for the West," *Foreign Affairs*, January/February 2009. Retrieved from http://www.foreignaffairs.com/articles/63714/roger-c-altman/the-great-crash-2008

AMIN, Samir, "Financial Collapse, Systemic Crisis?" World Forum for Alternatives, Caracas, October 2008. Retrieved from http://www.globalresearch.ca/index.php?context=va&aid=11099

BAHMANI, Sahar, "Understanding the Current Recession and Its Global Impact," *Gulf Coast Economics Association, 2009 Conference Proceedings*, Savannah Georgia, November 5, 2009, p. 11. Retrieved from http://gulfcoastecon.org/63312/62060.html

"Bailout is Law," *CNN Money*, October 4, 2008. Retrieved from http://money.cnn.com/2008/10/03/news/economy/house_friday_bailout/index.htm

BAJAJ, Vikas, "Home Prices fall for 10th Straight Month," *New York Times*, December 26, 2007. Retrieved from http://www.nytimes.com/2007/12/26/business/27home-web.html

BARR, Colin, "The $4 Trillion Housing Headache," *CNN Money*, May 27, 2009. Retrieved from http://money.cnn.com/2009/05/27/news/mortgage.overhang.fortune/index.htm

BERNANKE, Ben, "The Economic Outlook," Testimony before the Joint Economic Committee, October 20, 2005. Retrieved from http://www.house.gov/jec/hearings/testimony/109/10-20-05bernanke.pdf

BOGLE, John C., *The Battle for the Soul of Capitalism*, Yale University Press, New Haven, CT.

BROCKES, Emma, "He Told Us So," *Guardian*, January 24, 2009. Retrieved from http://www.guardian.co.uk/business/2009/jan/24/nouriel-roubini-credit-crunch

"Buffett Warns on Investment 'Time Bomb'," *BBC News*, March 4, 2003. Retrieved from http://news.bbc.co.uk/2/hi/2817995.stm

BUSH, George W., "President's Weekly Radio Address," August 6, 2005. In Robert J. Schiller, *The Subprime Solution: How Today's Global Financial Crisis Happened, and What to Do about It*, Princeton University Press, Princeton, NJ, p. 40.

"The Changing Narrative: How the News Media have Covered the Slowing Economy," *Analysis Report, Pew Research Center's Project for Excellence in Journalism*, August 8, 2001, p. 2. Retrieved from http://www.journalism.org/files/Economy%20 report.pdf

"Covering the Great Recession: Why Did Coverage of the Economy Decrease?" *Pew Research Center's Project for Excellence in Journalism*, October 5, 2009, p. 2. Retrieved from http://www. journalism.org/analysis_report/why_did_coverage_economy_ decrease

COY, Peter, MILLER, Rich, YOUNG, Lauren, and PALMERI, Christopher, "Is a Housing Bubble About to Burst?" *Bloomberg Businessweek*, July 19, 2004. Retrieved from http://www.businessweek.com/ magazine/content/04_29/b3892064_mz011.htm

COY, Peter, "What Good are Economists Anyway"? *Bloomberg Businessweek*, April 16, 2009. Retrieved from http://www. businessweek.com/magazine/content/09_17/b4128026997269.htm

"Criminal Fraud: Mortgage Fraud Scandal Brewing," *Real News*, May 13, 2009. Retrieved from http://therealnews.com/t2/index. php?option=com_content&task=view&id=31&Itemid=74& jumival=3708

"CSI: credit crunch," *Economist*, October 18, 2007. Retrieved from http:// www.economist.com/specialreports/displaystory.cfm?story_ id=9972489

"The Disappearing Dollar," *Economist*, December 2, 2004. Retrieved from http://www.economist.com/node/3446249

"Executive Summary," International Monetary Fund, January 2009. Retrieved from http://www.imf.org/external/pubs/ft/weo/2009/01/ pdf/exesum.pdf

"Existing-Home Sales Fall in 41 States," *Associated Press*, August 15, 2007. Retrieved from http://www.msnbc.msn.com/id/20279235/

"Declaration of the Summit on Financial Markets and the World Economy," Office of the Press Secretary, The White House, November 15, 2008. Retrieved from http://georgewbush-whitehouse.archives.gov/news/releases/2008/11/20081115-1.html

"Delinquencies and Foreclosures Increase in Latest MBA National Delinquency Survey," Mortgage Bankers Association, September 5, 2008. Retrieved from http://www.mbaa.org/NewsandMedia/PressCenter/64769.htm

"Delinquencies Continue to Climb in Latest MBA National Delinquency Survey," Mortgage Bankers Association, November 19, 2009. Retrieved from http://www.mbaa.org/NewsandMedia/PressCenter/71112.htm

FACKLER, Martin, "Trouble Without Borders," *New York Times*, October 23, 2008. Retrieved from http://query.nytimes.com/gst/fullpage.html?res=9B00E2DC103FF937A15753C1A96E9C8B63&ref=martinfackler

FIGLEWSKI, Stephen, SMITH, Roy C., and WALTER, Ingo, "Geithner's Plan for Derivatives," *Forbes*, May 18, 2009. Retrieved from http://www.forbes.com/2009/05/18/geithner-derivatives-plan-opinions-contributors-figlewski.html

GASSMAN, James K., "What to Learn from the Fall of Enron, a Firm that Fooled So Many," *International Herald Tribune*, December 10, 2001, p. 10.

GEITHNER, Timothy, "Reducing Systemic Risk in a Dynamic Financial System," Speech, Federal Reserve Bank of New York, June 9, 2008. Retrieved from http://www.newyorkfed.org/newsevents/speeches/2008/tfg080609.html

"Giant Pool of Money Wins Peabody," National Public Radio, This American Life, April 5, 2009. Retrieved from http://www.pri.org/stories/business/giant-pool-of-money.html

GLADWELL, Malcolm, The Tipping Point: How Little Things can Make a Big Difference, New York: Back Bay Books, 2002, p. 7.

GOEL, Suresh, Crisis Management: Master the Skills to Prevent Disasters, Global India Publications, New Delhi, India, p. 183.

GORDON, Robert, "Did Liberals Cause the Sub Prime Crisis?" *American Prospect*, April 7, 2008. Retrieved from http://prospect.org/article/did-liberals-cause-sub-prime-crisis

GREENSPAN, Alan, The Age of Turbulence: Adventures in a New World, Penguin Press HC, New York.

GULLAPALLI, Diya and ANAND, Shefalli, "Bailout of Money Funds Seems to Stanch Outflow," *Wall Street Journal*, September 20, 2008. Retrieved from http://online.wsj.com/article/SB122186683086958875.html?mod=article-outset-box

"A Helping Hand to Homeowners," *Economist*, October 23, 2008. Retrieved from http://www.economist.com/node/12470547?story_id=12470547

HOLMES, Steven A., "Fannie Mae Eases Credit to Aid Mortgage Lending," *New York Times*, September 30, 1999. Retrieved from http://www.nytimes.com/1999/09/30/business/fannie-mae-eases-credit-to-aid-mortgage-lending.html

"Home Equity Extraction: The Real Cost of 'Free Cash'," *Seeking Alpha*, April 25, 2007. Retrieved from http://seekingalpha.com/article/33336-home-equity-extraction-the-real-cost-of-free-cash

KEDROSKY, Paul, "How Enron Ran Out of Gas," *Wall Street Journal*, October 29, 2001, p. A22.

KRUGMAN, Paul, The Return of Depression Economics and the Crisis of 2008, New York, WW Norton, 2009, pp. 168–169.

KRUGMAN, Paul, "Revenge of the Glut," *New York Times*, March 1, 2009. Retrieved from http://www.nytimes.com/2009/03/02/opinion/02krugman.html?_r=1

KRUGMAN, Paul, "Reagan Did It," *New York Times*, May 31, 2009. Retrieved from http://www.nytimes.com/2009/06/01/opinion/01krugman.html

KRUGMAN, Paul, "Financial Reform 101," *New York Times*, April 1, 2010. Retrieved from http://www.nytimes.com/2010/04/02/opinion/02krugman.html?adxnnl=1&adxnnlx=1326978494-WOo7/m5tevHSi/qRiQCY5A

LABATON, Stephen, "Agency's '04 Rule Let Banks Pile Up New Debt," *New York Times*, October 3, 2008. Retrieved from http://www.nytimes.com/2008/10/03/business/03sec.html

LAMBERT, Richard, "Crashes, Bangs & Wallops," *Financial Times*, July 19, 2008. Retrieved from http://www.ft.com/cms/s/0/7173bb6a-552a-11dd-ae9c-000077b07658.html#axzz1jmaY8v3M

LANDER, Mark, "West is in Talks on Credit to Aid Poorer Nations," *New York Times*, October 24, 2008. Retrieved from http://www.nytimes.com/2008/10/24/business/worldbusiness/24iht-24emerge.17215442.html?pagewanted=all

LEE, Susan, "The Dismal Science: Enron's Success Story," *Wall Street Journal*, December 26, 2011, p. A11.

LEWIS, Michael, "The End of Wall Street's Boom," *Portfolio.com*, November 11, 2008. Retrieved from http://www.portfolio.com/news-markets/national-news/portfolio/2008/11/11/The-End-of-Wall-Streets-Boom

MANKIW, Gregory, "How to Avoid Recession? Let the Fed Work," *New York Times*, December 23, 2007. Retrieved from http://www.nytimes.com/2007/12/23/business/23view.html?ex=1356066000&en=3337604c8708710a&ei=5090&partner=rssuserland&emc=rss

MAX, Sara, "The Bubble Question, How Will Rising Interest Rates Affect Housing Prices?" *CNN Money*, July 27, 2004. Retrieved from http://money.cnn.com/2004/07/13/real_estate/buying_selling/risingrates/

MIHM, Stephen, "Dr. Doom," *New York Times*, August 15, 2008. Retrieved from http://www.nytimes.com/2008/08/17/magazine/17pessimist-t.html?pagewanted=all

"Minutes of the Federal Open Market Committee," Board of Governors of the Federal Reserve System, June 23-24, 2009, Washington D.C. Retrieved from http://www.federalreserve.gov/monetarypolicy/fomcminutes20090624.htm

MORGENSON, Gretchen, "A Bank Crisis Whodunit, With Laughs and Tears," *New York Times*, January 29, 2011. Retrieved from http://www.nytimes.com/2011/01/30/business/30gret.html

MORRIS, Charles, R., The Two Trillion Dollar Meltdown: Easy Money, High Rollers, and the Great Crash, New York, PublicAffairs, 2008.

"Multinational Arrangements," *World Academy Online*, June 2009. Retrieved from http://worldacademyonline.com/article/33/460/multinational_arrangements.html

NASTASE, Marian, CRETU, Alina Stefania, and STANEF, Roberta, "Effects of Global Financial Crisis," *Review of International Comparative Management*, 10, 2009, p. 695. Retrieved from www.rmci.ase.ro/no10vol4/Vol10_No4_Article9.pdf

NOCERA, Joe, "As Credit Crisis Spiraled, Alarm Led to Action," *New York Times*, October 1, 2008. Retrieved from http://www.nytimes.com/2008/10/02/business/02crisis.html

NORRIS, Floyd, "Another Crisis, Another Guarantee," *New York Times*, November 24, 2008. Retrieved from http://www.nytimes.com/2008/11/25/business/25assess.html?hp

"Open Market Operations," Board of Governors of the Federal Reserve System, January 26, 2010. Retrieved from http://www.federalreserve.gov/monetarypolicy/openmarket.htm

POWELL, Michael, "Crises in Japan Ripple Across the Global Economy," New York Times, March 20, 2011. Retrieved from http://www.nytimes.com/2011/03/21/business/global/21econ.html?pagewanted=all

RealtyTrac Staff, "U.S. Foreclosure Activity Increases 75 Percent in 2007," RealtyTrac, January 30, 2008. Retrieved from http://www. realtytrac.com/content/press-releases/us-foreclosure-activity- increases-75-percent-in-2007-3604?accnt=64847

REILLY, David, "Banks' Hidden Junk Menaces $1 Trillion Purge," *Bloomberg*, March 25, 2009. Retrieved from http://www. bloomberg.com/apps/news?pid=newsarchive&sid=akv_ p6LBNIdw&refer=home

SAKOLSKI, Aaron M., *The Great American Land Bubble: The Amazing Story of Land Grabbing, Speculations, and Booms from Colonial Days to the Present Time*, Johnson Reprint Corp., New York, NY.

SALMON, Felix, "Recipe for Disaster: The Formula That Killed Wall Street," *Wired Magazine*, February 23, 2009. Retrieved from http://www.wired.com/techbiz/it/magazine/17-03/wp_ quant?currentPage=all

SCHILLER, Robert J., The Subprime Solution: How Today's Global Financial Crisis Happened, and What to Do about It, Princeton University Press, Princeton, NJ, p. 47.

SOROS, George, "The Worst Market Crisis in 60 Years," *Financial Times*, January 22, 2008, Retrieved from http://www. ft.com/cms/s/0/24f73610-c91e-11dc-9807-000077b07658. html#axzz1jtkkWJ7H

"Spending Boosted by Home Equity Loans: Greenspan," *Reuters*, April 23, 2007. Retrieved from http://www.reuters.com/article/2007/04/23/ us-usa-greenspan-equity-idUSN2330071920070423

STEVERMAN, Ben and BOGOSLAW, David, "The Financial Crisis Blame Game," *Bloomberg Businessweek*, October 18, 2008. Retrieved from http://www.businessweek.com/investor/content/oct2008/ pi20081017_950382.htm?chan=top+news_top+news+index+- +temp_top+story

"Ted Spread," *Bloomberg*. Retrieved from http://www.bloomberg.com/ quote/!TEDSP:IND

TIMRAOS, Nick and BRAY, Chad, "SEC Brings Crisis-Era Suits," *Wall Street Journal*, December 17, 2011. Retrieved from http://online.wsj.com/article/SB1000142405297020373330457710231095578 0788.html

TRUMP, Donald, CBS Early Show, March 6, 2009.

"United States GDP Growth Rate," *Trading Economics*. Retrieved from http://www.tradingeconomics.com/united-states/gdp-growth

"U.S., European Bank Writedowns, Credit Losses," *Reuters*, November 5, 2009. Retrieved from http://www.reuters.com/article/2009/11/05/banks-writedowns-losses-idCNL554155620091105?rpc=44

WALLISON, Peter J., "The True Origins of This Financial Crisis," *American Spectator*, February 6, 2009. Retrieved from http://spectator.org/archives/2009/02/06/the-true-origins-of-this-finan

"Will Subprime Mess Ripple through the Economy?" *MSNBC*, March 13, 2007. Retrieved from http://www.msnbc.msn.com/id/17584725#.Txf6qmOonus

WILSON, James and KELLING, George, "Broken Windows. The Police and Neighbourhood Safety," *Atlantic Magazine*, 3, 1982.

WOLF, Martin, "Japan's Lessons for a World of Balance-Sheet Deflation," *Financial Times*, February 17, 2009. Retrieved from http://www.ft.com/intl/cms/s/0/774c0920-fd1d-11dd-a103-000077b07658.html#axzz1jtkkWJ7H

WOLF, Martin, "Reform of Regulation has to Start by Altering Incentives," *Financial Times*, June 23, 2009. Retrieved from http://www.ft.com/intl/cms/s/0/095722f6-6028-11de-a09b-00144feabdc0.html#axzz1jtkkWJ7H

ZUCKERMAN, Mortimer, "The Economy Is Even Worse Than You Think," *Wall Street Journal Opinion section*, July 14, 2009. Retrieved from http://online.wsj.com/article/SB124753066246235811.html

Chapter 5: Climate Change

ADAM, David, "Climate Change Sceptics and Lobbyists put World at Risk, says Top Adviser," *Guardian*, November 22, 2009. Retrieved from http://www.guardian.co.uk/environment/2009/nov/22/climate-change-emissions-scientist-watson

"Administration To Deny Global Warming," *Rolling Stone*, June 20, 2007. Retrieved from http://www.desmogblog.com/sites/beta.desmogblog.com/files/The%20Secret%20Campaign%20of%20President%20Bush%20rolling%20stone.pdf

"An Increase in GOP Doubt About Global Warming Deepens Partisan Divide," *Pew Research Center for the People and the Press*, May 8, 2008. Retrieved from http://pewresearch.org/pubs/828/global-warming

BELL, Larry, "Hot Sensations vs. Cold Facts," Forbes, January 28, 2011. Retrieved from http://www.forbes.com/sites/larrybell/2011/01/28/hot-sensations-vs-cold-facts-3/

BELL, Larry, *Climate of Corruption: Politics and Power Behind the Global Warming Hoax*, Austin, TX, Greenleaf Book Group, 2011, p. 84.

BODEN, Thomas A., MARLAND, Gregg, and ANDRES, Robert J., "Global, Regional, and National Fossil-Fuel CO_2 Emissions," Carbon Dioxide Information Analysis Center, Oak Ridge, TN, 2006. Retrieved from http://cdiac.ornl.gov/trends/emis/overview_2006.html

BORENSTEIN, Seth, "Obama Science Advisers Grilled over Hacked E-mails," *Breitbart*, December 2, 2009. Retrieved from http://www.breitbart.com/article.php?id=D9CBFB901

BOYCOFF, Maxwell T. and BOYCOFF, Jules M., "Balance as Bias: Global Warming and the US Prestige Press," *Global Environmental Change*, 14, 2004, pp. 125–136.

BRECKE, Peter, "Violent Conflicts 1400 A. D. to the Present in Different Regions of the World," Annual Meeting of the Peace Science Society (International), Ann Arbor, MI, October 8–10, 1999.

BRODER, John M., "Climate Deal Likely to Bear Big Price Tag," *New York Times*, December 8, 2009. Retrieved from http://www.nytimes.com/2009/12/09/science/earth/09cost.html?pagewanted=all

BROHAN, P., KENNEDY, J J., HARRIS, I., TETT, S. F. B., and JONES, P. D., "Uncertainty Estimates in Regional and Global Observed Temperature Changes: A New Data Set from 1850," *Journal of Geophysical Research Atmospheres*, 111, 2006, D12106.

BUHAUG, Halvard, *Second IMO GHG Study 2009*, London, UK, International Maritime Organization (IMO), 2009. Retrieved from http://www.imo.org/blast/blastDataHelper.asp?data_id=27795&filename=GHGStudyFINAL.pdf

CAMPBELL, Duncan, "White House Cuts Global Warming from Report," *Guardian*, June 20, 2003. Retrieved from http://www.guardian.co.uk/environment/2003/jun/20/climatechange.climatechangeenvironment

CHARNEY, Jule G., "Carbon Dioxide and Climate: A Scientific Assessment," National Academy of Sciences Summer Studies Center, Washington, D.C., National Academy of Sciences, 1979. Retrieved from www.atmos.ucla.edu/~brianpm/download/charney_report.pdf

"Climate Change Fight 'Can't Wait'," *BBC*, October 31, 2006. Retrieved from http://news.bbc.co.uk/2/hi/business/6096084.stm

"Climate Chaos: Bush's Climate of Fear," *BBC*, June 1, 2006. Retrieved from http://news.bbc.co.uk/2/hi/programmes/panorama/5005994.stm

COILE, Zachary, "How the White House worked to scuttle California's climate law," San Francisco Chronicle, September 25, 2007. Retrieved from http://articles.sfgate.com/2007-09-25/news/17261302_1_auto-emissions-greenhouse-gases-e-mails

Committee on the Science of Climate Change, National Research Council, *Climate Change Science: An Analysis of Some Key Questions*, Washington, D.C., National Academies Press, 2001.

"Copenhagen Climate Accord: Key Issues," *BBC*, December 19, 2009. Retrieved from http://news.bbc.co.uk/2/hi/8422186.stm

COWIE, Jonathan, *Climate and Human Change: Disaster or Opportunity?* New York, Parthenon, 1998.

DEN ELZEN, Michel and HOHNE, Niklas, "Reductions of Greenhouse Gas Emissions in Annex I and non-Annex I Countries for Meeting Concentration Stabilisation Targets," *Climatic Change*, 91, 2008, pp. 247–274.

DEN ELZEN, Michel G. J., VAN VUUREN, Detlef P., and VAN VLIET, Jasper, "Postponing Emission Reductions from 2020 to 2030 Increases Climate Risks and Long-Term Costs," Climatic Change, 99, 2010, pp. 313–320.

DICKINSON, Tim, "The Secret Campaign of President Bush's

DONEY, Scott C., FABRY, Victoria J., FEELY, Richard A., and KLEYPAS, Joan A., "Ocean Acidification: The Other CO_2 Problem," *Annual Reviews*, 1, 2009, pp. 169–192.

EILPERIN, Juliet, "Climate Researchers Feeling Heat From White House," *Washington Post*, April 6, 2006. Retrieved from http://www.washingtonpost.com/wp-dyn/content/article/2006/04/05/AR2006040502150_pf.html

Emissions Scenarios, Intergovernmental Panel on Climate Change (IPCC), Nebojsa Nakicenovic and Rob Swart (eds.), Cambridge and New York, Cambridge University Press, 2000.

European Commission Joint Research Centre (JRC)/Netherlands Environmental Assessment Agency (PBL), Emission Database for Global Atmospheric Research (EDGAR), 2009, release version 4.0. Retrieved from http://edgar.jrc.ec.europa.eu/overview.php?v=40

FARKAS, Tamás, *The Investor's Guide to the Energy Revolution*, Raleigh, NC., Lulu.com, 2008, p. 234.

FERGUSON, R. Brian, *Warfare, Culture, and Environment*, Orlando, FL, Academic, 1984.

FRIEDMAN, Thomas L., "The Earth is Full," *New York Times*, June 7, 2011. Retrieved from http://www.nytimes.com/2011/06/08/opinion/08friedman.html

FURUYA, Jun, KOBAYASHI, Shintaro, and MEYER, Seth D., "Economic Impacts of Climate Change on Global Food Supply and Demand," *Japan Agriculture Research*, 39, 2005, pp. 121–134.

GALLOWAY, Patrick R., "Long-Term Fluctuations in Climate and Population in the Preindustrial Era," *Population and Development Review*, 12, 1986, pp. 1–24.

GILDING, Paul, *The Great Disruption: Why the Climate Crisis Will Bring On the End of Shopping and the Birth of a New World*, New York, NY, Bloomsbury Press, 2011.

GILLIS, Justin, "A Scientist, His Work and a Climate Reckoning," *New York Times*, December 21, 2010. Retrieved from http://www.nytimes.com/2010/12/22/science/earth/22carbon.html

GOLDSTEIN, Natalie and COOK, Kerry Harrison, *Global Warming*, New York, Checkmark Books, 2010, p. 164.

"Groups Say Scientists Pressured On Warming," *CBS News*, February 11, 2009. Retrieved from http://www.cbsnews.com/stories/2007/01/30/politics/main2413400.shtml

HAMILTON, Tyler, "Fresh Alarm over Global Warming," *Toronto Star*, January 1, 2007. Retrieved from http://www.thestar.com/article/166819

HIRSCH, Robert L., report, Science Applications International Corporation.

HOEGH-GULDBERG, Ove and BRUNO, John F., "The Impact of Climate Change on the World's Marine Ecosystems," *Science*, 18, 2010, pp. 1523-1528.

HOGGAN, James and LITTLEMORE, Richard D., *Climate Cover-up: The Crusade to Deny Global Warming*, Vancouver, Canada, Greystone Books, 2009, p. 19.

HOMER-DIXON, Thomas F., "Environmental Scarcities and Violent Conflict: Evidence from Cases," *International Security*, 19, 1994, pp. 5–40.

"IPCC Second Assessment, Climate Change 1995," Intergovernmental Panel on Climate Change, December 1995. Retrieved from http://www.ipcc.ch/pdf/climate-changes-1995/ipcc-2nd-assessment/2nd-assessment-en.pdf

"IPCC Third Assessment, Climate Change 1995," Intergovernmental Panel on Climate Change (IPCC), 2001. Retrieved from http://www.ipcc.ch/ipccreports/tar/index.htm

JACOBS, Andrew, "China Issues Warning on Climate and Growth," *New York Times*, February 28, 2011. Retrieved from http://www.nytimes.com/2011/03/01/world/asia/01beijing.html

JOHNSON, Douglas L. and GOULD, Harvey A., "The Effect of Climate Fluctuations on Human Populations: A Case Study of Mesopotamian Society," in *Climate and Development*, Asit K. BISWAS (ed.), Dublin, Tycooly International Limited, 1984, pp. 117–138.

KAHN, Matthew E., *Climatopolis: How our Cities Will Thrive in the Hotter Future*, New York, Basic Books, p. 6.

KLUGER, Jeffrey, "A Climate of Despair," Time, April 1, 2001. Retrieved from http://www.time.com/time/magazine/article/0,9171,104596,00.html

KRUGMAN, Paul, "Boiling the Frog," *New York Times*, July 12, 2009. Retrieved from http://www.nytimes.com/2009/07/13/opinion/13krugman.html

LANDERER, Felix W., JUNGCLAUS, Johann H., and MAROTZKE, Jochem, "Regional Dynamic and Steric Sea Level Change in Response to the IPCC-A1B Scenario," Journal of Physical Oceanography, 37, 2006, pp. 296–312.

LEE, Harry F., FOK, Lincoln, and ZHANG, David D., "Climatic Change and Chinese Population Growth Dynamics over the Last Millennium," *Climatic Change*, 88, 2007, pp. 131–156.

LOBELL David B., and FIELD, Christopher B., "Global Scale Climate–Crop Yield Relationships and the Impacts of Recent Warming," *Environmental Research Letters*, 2, 2007.

MANN, Michael E. and JONES, Philip D., "Global Surface Temperatures over the Past Two Millennia," *Geophysical Research Letters*, 30, 2003.

MASLOW, Abraham Harold, *Motivation and Personality*, New York, Harper & Row, 1970.

McKIE, Robin "Why Channel 4 has got it wrong over climate change." London: Guardian Unlimited. 4 March 2007. Retrieved from http://environment.guardian.co.uk/climatechange/story/0,,2026125,00.html.

MCNEIL, Ben I. and MATEAR, Richard J., "Southern Ocean Acidification: A Tipping Point at 450-ppm Atmospheric CO_2," *Proceedings of the National Academy of Sciences*, 105, 2008, pp. 18860–18864.

MEINSHAUSEN, Malte, HARE, Bill, VAN VUUREN, Detlef P., DEN ELZEN, Michel, and SWART, Rob, "Multi-Gas Emission Pathways to Meet Climate Targets," *Climate Change*, 75, 2006, pp. 151–194.

MEINSHAUSEN, Malte, RAPER, Sarah, and WIGLEY, Tom, "Emulating IPCC AR4 Atmosphere-Ocean and Carbon Cycle Models for Projecting Global-Mean, Hemispheric and Land/Ocean Temperatures: MAGICC 6.0," *Atmospheric Chemistry and Physics*, 8, 2008, pp. 6153–6272.

MEINSHAUSEN, Malke and RAPER, Sarah, *The Rising Effect of Aviation on Climate; Project Report OMEGA—Aviation in a Sustainable World*, Manchester, UK, Manchester Metropolitan University, 2009.

MEINSHAUSEN, Malte, MEINSHAUSEN, Nicolai, HARE, William, RAPER, Sarah C. B., FRIELER, Katja, KNUTTI, Reto, FRAME, David J., and ALLEN, Myles R., "Greenhouse-Gas Emission Targets for Limiting Global Warming to 2 °C," *Nature*, 458, 2009, pp. 1158–1162.

MONBIOT, George (30 January 2007). "Don't be fooled by Bush's defection: his cures are another form of denial." London: The Guardian. http://www.guardian.co.uk/commentisfree/story/0,,2001694,00.html.

MONBIOT, George (2007-03-13). "Don't let truth stand in the way of a red-hot debunking of climate change." London: The Guardian. http://environment.guardian.co.uk/climatechange/story/0,,2032572,00.html.

MOONEY, Chris, *The Republican War on Science*, New York, Basic Books, 2005.

NABEL, Jane, MACEY, Kirsten, and CHEN, Claudine, "PRIMAP Reference Data for LULUCF Accounting," Potsdam Real-time Integrated Model for probabilistic Assessment of emissions Paths (PRIMAP). Retrieved from www.primap.org

NABEL, Julia, ROGELJ, Joeri, CHEN, Claudine M., MARKMANN, Kathleen, GUTZMANN, David J., and MEINSHAUSEN, Malte, "Decision Support for International Climate Policy—the PRIMAP Emission Module," *Environmental Modelling and Software*, 26, 2011, pp. 1419–1433.

"NASA Looks at Seal Level Rise, Hurricane Risks to New York City," NASA Goddard Institute for Space Studies, October 24, 2006. Retrieved from http://www.giss.nasa.gov/research/news/20061024/

PETERS, Sandra L., MALCOLM, Jay R., and ZIMMERMAN, Barbara L., "Effects of Selective Logging on Bat Communities in the Southeastern Amazon," *Conservation Biology*, 20, 2006, pp. 1410–1421.

PORTER, Henry, "Fiddling as the Planet Burns," The Observer, June 19, 2005. Retrieved from http://www.guardian.co.uk/politics/2005/jun/19/greenpolitics.climatechange

"President Bush Discusses Global Climate Change," Office of the Press Secretary, The White House, June 11, 2001. Retrieved from http://georgewbush-whitehouse.archives.gov/news/releases/2001/06/20010611-2.html

PULLELLA, Philip, "Pope Urges, Save the Planet Before it's too Late," *Reuters*, September 2, 2007. Retrieved from http://www.enn.com/top_stories/commentary/22598/print

PURVIS, Andrew, "Heroes of the Environment: Angela Merkel," *Time*, October 17, 2007. Retrieved from http://www.time.com/time/specials/2007/article/0,28804,1663317_1663319_1669897,00.html

REVELLE, Roger and SEUSS, Hans, "Carbon Dioxide Exchange between Atmosphere and Ocean and the Question of an Increase of Atmospheric CO_2 during the Past Decades," *Tellus*, 9, 1957, pp. 18–27.

REVKIN, Andrew C. and BRODER, John M., "In Face of Skeptics, Experts Affirm Climate Peril," *New York Times*, December 6, 2009. Retrieved from http://www.nytimes.com/2009/12/07/science/earth/07climate.html

REVKIN, Andrew C. and BRODER, John M., "Facing Skeptics, Climate Experts Sure of Peril," *New York Times*, December 7, 2009, pp. A1, A8.

RIAHI, Keywan, GRUBLER, Amulf, and NAKICENOVIC, Nebojsa, "Scenarios of Long-Term Socio-Economic and Environmental Development under Climate Stabilization," *Technological Forecasting and Social Change*, 74, 2007, pp. 887–935.

ROBELIUS, Frederik, Doctoral dissertation, University of Uppsala, Sweden.

ROGELJ, Joeri, NABEL, Julia, CHEN, Claudine, HARE, William, MARKMANN, Kathleen, MEINSHAUSEN, Malte, SCHAEFFER, Michiel, MACEY, Kirsten, and HOHNE, Niklas, "Copenhagen Accord pledges are paltry," *Nature*, 464, 2010, pp. 1126–1128.

"Rudd Ratifies Kyoto," *The Age*, December 3, 2007. Retrieved from http://www.theage.com.au/news/national/rudd-ratifies-kyoto/2007/12/03/1196530553722.html

SAAD, Lydia, "Increased Number Think Global Warming Is 'Exaggerated',"
    Gallup, March 11, 2009. Retrieved from http://www.gallup.com/
    poll/116590/increased-number-think-global-warming-exaggerated.
    aspx

SEAGER, Richard, GRAHAM, Nicholas, and HERWEIJER, Celine,
    "Blueprints for Medieval Hydroclimate," *Quaternay Science
    Reviews*, 26, 2007, pp. 19–21.

SILVERMAN, Jacob, LAZAR, Boaz, CAO, Long, CALDERA, Ken, and
    EREZ, Jonathan, "Coral Reefs May Start Dissolving When
    Atmospheric CO_2 Doubles," *Geophysical Research Letters*, 36,
    2009, L05606.

SMITH, Steven J. and WIGLEY, T.M.L., "Multi-Gas Forcing Stabilisation
    with the MiniCAM," *Energy Journal* (special issue 3), 2006,
    pp. 373–391.

"Smoke, Mirrors & Hot Air: How ExxonMobil uses Big Tobacco's Tactics
    to Manufacture Uncertainty on Climate Science," Cambridge, MA,
    Union of Concerned Scientists, 2007. Retrieved from http://www.
    ucsusa.org/assets/documents/global_warming/exxon_report.pdf

STEINACHER, Marco, JOOS, Fortunat, FROLICHER, Thomas L.,
    PLATTNER, Gian Kasper, and DONEY, Scott, "Imminent Ocean
    Acidification in the Arctic Projected with the NCAR Global
    Coupled Carbon Cycle-Climate Model," *Biogeosciences*, 6, 2009,
    pp. 515–533.

STERN, Nicholas, *Stern Review on the Economics of Climate Change*,
    Her Majesty's Treasury, 2006. Retrieved from http://webarchive.
    nationalarchives.gov.uk/+/http://www.hm-treasury.gov.uk/stern_
    review_report.htm

STERN, Nicholas, *Deciding Our Future in Copenhagen: Will the World
    Rise to the Challenge of Climate Change?* public lecture, London,
    UK, Grantham Research Institute for Climate Change and the
    Environment, 2009.

SUHRKE, Astri, "Environmental Degradation, Migration, and the Potential for Violent Conflict," in *Conflict and the Environment*, Nils Petter GLEDITSCH (ed.), The Netherlands, Kluwer, Dordrecht, 1997, pp. 255–272.

THOMPSON, Andrea, "Timeline: Earth's Precarious Future," *Live Science*, January 11, 2008. Retrieved from http://www.livescience.com/1433-timeline-earth-precarious-future.html

U.S. National Assessment, U.S. Global Change Research Program, "Climate Change Impacts on the United States: The Potential Consequences of Climate Variability and Change," 2001, Cambridge, UK, Cambridge University Press.

United Nations Department of Economic and Social Affairs/Population Division (UNDES), "World Population Prospects, The 2008 Revision," 2008. Retrieved from http://www.un.org/esa/population/publications/wpp2008/wpp2008_highlights.pdf

United Nations Environment Program (UNEP) Chief Scientists Office, "How Close Are We to the Two Degree Limit?" UNEP Governing Council Meeting & Global Ministerial Environment Forum, February 24-26, 2010, Bali, Indonesia.

United Nations Framework Convention on Climate Change (UNFCCC), June 28, 2002, FCCC/INFORMAL/84. Retrieved from http://unfccc.int/resource/docs/convkp/conveng.pdf

United Nations Framework Convention on Climate Change (UNFCCC) Secretariat, "Negotiating Text. Note by the Secretariat," Bonn, Switzerland, United Nations Office, August 13, 2010, FCCC/AWGLCA/2010/14.

United States. "Restoring the Quality of Our Environment," Environmental Pollution Panel, President's Science Advisory Committee, Washington, D.C., White House, 1965.

"U.S. and World Population Clocks," U.S. Census Bureau. Retrieved from http://www.census.gov/main/www/popclock.html

VAN VUUREN, Detlef P., LUCAS, Paul L., and HILDERINK, Henk, "Downscaling Drivers of Global Environmental Change: Enabling Use of Global SRES Scenarios at the National and Grid Levels," *Global Environmental Change*, 17, 2007, pp. 114–130.

VERON, John E. N., HOEGH-GULDBERG, Ove, M. LENTON, Tim, LOUGH, Janice M., OBURA, David O., PEARCE-KELLY, Paul, SHEPPARD, Charles R.C., SPALDING, Mark, STAFFORD-SMITH, M.G., and D. ROGERS, Alex, "The Coral Reef Crisis: The Critical Importance of <350 ppm CO_2," *Marine Pollution Bulletin*, 58, 2009, pp. 1428–1436.

VICTOR, David G., *Climate Change: Debating America's Policy Options*, New York, Council on Foreign Relations Press, 2004, pp. 143–144.

WATSON, Robert. "The public has been swindled." *The Guardian*, 28 July 2008

WEBSTER, David, "Warfare and the Evolution of the State: A Reconsideration," *American Antiquity*, 40, 1975, pp. 464–470.

"White House 'Eviscerated' CDC Testimony Regarding Climate Change and Health," *Associated Press*, October 24, 2007. Retrieved from http://www.commondreams.org/archive/2007/10/24/4772

WILLIAMS, John W., JACKSON, Stephen T., and KUTZBACH, John E., "Projected distributions of novel and disappearing climates by 2100 AD," *Proceedings of the National Academy of Sciences*, 104, 2007, pp. 5738–5742.

"World Energy Outlook 2009," International Energy Agency, Paris, France, 2009. Retrieved from http://www.worldenergyoutlook.org/docs/weo2009/WEO2009_es_english.pdf

"World Population in 2300 could Stabilize at 9 Billion, UN Estimates," United Nations News Centre, November 4, 2004. Retrieved from http://www.un.org/apps/news/story.asp?NewsID=12439

ZABARENKO, Deborah, "US Climate Scientists Allege White House Pressure," Reuters, January 30, 2007. Retrieved from http://www.commondreams.org/headlines07/0130-10

ZELLER Jr., Tom, "And in This Corner, Climate Contrarians," *New York Times*, December 9, 2009. Retrieved from http://www.nytimes.com/2009/12/10/science/earth/10skeptics.html

ZHANG, David D., JIM, C. Y., LIN, George C. S., HE, Yuan-Qing, WANG, James J., and LEE, Harry F., "Climate Change, Wars and Dynastic Cycles in China over the Last Millennium," *Climatic Change*, 76, 2006, pp. 459–477.

ZHANG, David D., BRECKE, Peter, LEE, Harry F., HE, Yuan-Qing, and ZHANG, Jane, "Global Climate Change, War, and Population Decline in Recent Human History," *Proceedings of the National Academy of Sciences*, 104, 2007, pp. 19214–19219.

ZHANG, David D., ZHANG, Jane, LEE, Harry F., and HE, Yuan-Qing, "Climate Change and War Frequency in Eastern China over the Last Millennium," *Human Ecology*, 35, 2007, pp. 403–414.

Part III: References

Chapter 6: Comparisons and Historical Context

Dean, Cornelia, "Group Urges Research Into Aggressive Efforts to Fight Climate Change," *New York Times*, October 4. 2011. Retrieved from http://www.nytimes.com/2011/10/04/science/earth/04climate. html.

"The Environment in the News," United Nations Environment Programme, July 22, 2011. Retrieved from www.unep.org/cpi/ briefs/2011July22.doc, p. 9.

Chapter 7: Conclusions

"The Environment in the News," United Nations Environment Programme, July 22, 2011. Retrieved from www.unep.org/cpi/ briefs/2011July22.doc, p. 9.

Cornelia DEAN, "Group Urges Research Into Aggressive Efforts to Fight Climate Change," New York Times, October 4. 2011. Retrieved from http://www.nytimes.com/2011/10/04/science/earth/04climate. html.

READ MORE PUBLICATIONS
FROM BLUE MATRIX

Available Now:

Climate Change: Tipping Points Predicting Our Global Future

Climate Change Tipping Points are in many ways, influenced by media. This just released, ground-breaking book by Dr. Philip Gordon, PhD, details the Climate Change issue which is analyzed on the basis of Tipping Point *Attributes* and *Characteristics. Media Contagiousness* and *Stickiness* influencing the *epidemic* and (soon) *dramatic moment* in time, which is characterized by the *phenomenon* of a Tipping Point.

As the issue of Climate Change is still evolving, Dr. Gordon provides a methodology and parameters for analyzing and predicting our planets' most pressing global issue. Our species challenge is to harness the media global consciousness and refocus our global agendas and priorities before it's too late.

ISBN-10: 1481069225
ISBN-13: 978-1481069229

 http://www.amazon.com/Climate-Change-PhD-Philip-Gordon/dp/1481069225

 http://www.amazon.com/Climate-Change-Tipping-Points-ebook/dp/B00AFWLHDC

Coming soon on Smashwords!

Media Tipping Points

Tipping Points as evidenced in *global events* are, in many ways, influenced by media. This insightful book by Dr. Philip Gordon, PhD, details three case studies which were selected on the basis of common Tipping Point *attributes*: they each involved *media contagiousness* and *stickiness* during their development, and, each arrived at a moment in time, which could be characterized by the *phenomenon* of *Tipping Points*.

The first case study, the Presidential Campaign of Barack Obama was chosen in order to examine a narrower scope and timeframe for the application of the analysis, in contrast to the second case study, the International Financial Crisis of 2007-2010, which involved broader and more complex issues, and finally, the last case study, Climate Change, is included as consideration in this book as the research reveals critical relationships between Media Impact and the *Tipping Points* of this upcoming Global Event.

As the issue of Climate Change is still evolving, Gordon provides a methodology and parameters for analyzing and predicting the relative media impact on a global scale.

ISBN-10: 098476383X
ISBN-13: 978-0984763832

 http://www.amazon.com/Media-Tipping-Points-Analyzing-Predicting/dp/098476383X/

 eBook version coming soon on Smashwords.com

Obama Presidential Campaign: Tipping Point Media Analysis and Influence

Dr. Philip Gordon, PhD

On October 28, nine days before the 2012 US presidential elections, the east coast was devastated by the *"perfect storm"*: Hurricane Sandy. More than 100 people died, 50 billion USD in damages, communities destroyed and millions without services...

As in 2008, with the International Financial Crisis, Barack Obama again benefited from the *"Something else"* of global proportions and a media convulsion that guaranteed his epidemic Tipping Point victory.

Presidential Campaigns are in many ways, influenced by media. This just released, ground-breaking book by Dr. Philip Gordon, PhD, details the Obama Presidential Campaign: 2007–08, which is analyzed on the basis of *Tipping Point Theory Attributes* and *Characteristics*.

Dr. Gordon provides a methodology and parameters for analyzing and predicting presidential campaigns and other global issues.

ISBN-10: 1481129988
ISBN-13: 978-1481129985

 http://www.amazon.com/Obama-Presidential-Campaign-Analysis-Influence/dp/1481129988

 http://www.amazon.com/Obama-Presidential-Campaign-Tipping-ebook/dp/B00AORRGKQ

Principles and Practices of Lighting Design: The Art of Lighting Composition

A complete handbook on Lighting Design with both Artistic and Technical approaches for the beginning to advanced lighting designer.

Dr. Philip Gordon, PhD, LC, NCQLD has written a *"must have"* handbook that includes applications and case studies as well as updated product advances, specifications, resources and guides. Suitable for educational programs, professionals and related design fields. Also includes theory and resources for individuals interested in the future of the lighting design profession. Philip has combined several types of lighting books (theory, application, case study and resource) into one volume. A perfect resource for any design reference.

ISBN-10: 0615471633
ISBN-13: 978-0615471631

 http://www.amazon.com/Principles-Practices-Lighting-Design-Composition/dp/0615471633/

 http://www.smashwords.com/books/view/98230

Over 65: Writing for Pleasure and Profit!

Over 65: Writing for Pleasure and Profit! shows you how to be a successful, published writer and... It's simple with our proven method!

From ecommerce, to short stories, to TV and radio commercials, to publishing your own book, the Writing for Pleasure and Profit book gives you the tips to finding jobs and Getting Published!

ISBN-10: 0984763880
ISBN-13: 978-0984763887

 http://www.amazon.com/Over-65-Writing-Pleasure-Profit/dp/0984763880

 http://www.amazon.com/Over-65-Writing-Pleasure-ebook/dp/B009ZE7PIC

The International Financial Crisis: 2007-2010 Tipping Points

Qualitative Analysis, Comparisons and Predictions

The International Financial Crisis of 2007–2010 was, in many ways, influenced by media. This just released, ground-breaking book by Dr. Philip Gordon, PhD, details the International Financial Crisis: 2007–2010 which is analyzed on the basis of Tipping Point *Attributes* and *Characteristics*.

Dr. Gordon provides a methodology and parameters for analyzing and predicting our planets' most pressing global issue

ISBN-10: 1481244272
ISBN-13: 978-1481244275

 http://www.amazon.com/International-Financial-Crisis-2007-2010-Tipping/dp/1481244272

 http://www.amazon.com/International-Financial-Crisis-2007-2010-ebook/dp/B00ASJWAB0

Music CD's and DVD's

Available from Blue Matrix Publications
http://bluematrix.org/

www.MDG-Consultants.com

www.ingramcontent.com/pod-product-compliance
Lightning Source LLC
Chambersburg PA
CBHW081119170526
45165CB00008B/2498